FIRE DEPARTMENT INTERVIEW TACTICS

FIRE DEPARTMENT INTERVIEW TACTICS

GENE MAHONEY

THOMSON
DELMAR LEARNING

Australia Brazil Canada Mexico Singapore Spain United Kingdom United States

Fire Department Interview Tactics
Gene Mahoney

Vice President, Technology and Trades ABU:
David Garza

Director of Learning Solutions:
Sandy Clark

Acquisitions Editor:
Alison Weintraub

Product Manager:
Jennifer A. Thompson

Marketing Director:
Deborah S. Yarnell

Channel Manager:
Erin Coffin

Marketing Specialist:
Penelope Crosby

Senior Production Manager:
Larry Main

Content Project Manager:
Jennifer Hanley

Technology Project Specialist:
Linda Verde

Editorial Assistant:
Maria Conto

Printed in the United States of America

1 2 3 4 5 XX 08 07 06

For more information contact Thomson Delmar Learning
Executive Woods
5 Maxwell Drive, PO Box 8007,
Clifton Park, NY 12065-8007
Or find us on the World Wide Web at
www.delmarlearning.com

Library of Congress Cataloging-in-Publication Data

Mahoney, Gene, 1923–

Fire department interview tactics/ Gene Mahoney.

p. cm.
Includes index.
ISBN: 1-4180-3004-X

1. Fire prevention--Vocational guidance.
2. Employment interviewing. 3. Fire fighters--Employment. 4. Employment portfolios. I. Title.
TH9119.M335 2006
363.37023--dc22

2006013116

CONTENTS

PREFACE

Intent of This Book

Although this book has been prepared for use in a fire department environment for entrance and promotional interviews, the principles outlined within should prove valuable in preparing for an oral interview for any occupation. Interviewers are basically seeking the same intangibles, regardless of whether the person being interviewed will work for a fire department, police department, water department, or any other civil service department. Industrial applicants are also screened for these same intangibles. Of course, the psychological make-up of an individual subjected to stress, whose decisions may affect the loss or preservation of life and property, should be different from another whose work environment is peaceful and relatively free from stress. However, the oral interview principles and practices for any work environment will be basically the same, with only the type and depth of the questions changing.

The fire department environment has been used in this manual because no other occupation provides a setting in which people are required to work for long periods in situations demanding close harmony among peers, yet at a moment's notice individuals can be transferred from a serene setting to one imposing maximum physical and mental stress—situations in which a wrong decision can result in the loss of life.

Because of these rapidly changing environments and demands, the scope of questioning in a fire department oral interview approaches the maximum. A person capable of performing satisfactorily in interviews demanding this breadth of ability should be able to perform capably in any interview environment.

How to Use This Book

This book is a training manual. It requires the active participation of the reader in order to gain the maximum benefit from its use. Some people will use this training manual as a textbook. Without a doubt, they will benefit from doing so. However, if you are sincere in your desire to improve your performance during an oral interview, you will use it as a training manual. Such use requires that you complete all the assignments included in each chapter, and you follow the advice given throughout the manual. Keep in mind that this manual has been prepared and presented with one thought in mind—to help you obtain the position you are seeking.

This training manual includes the entire scope of oral interviews. It begins by providing the reader with a thorough background on the examination process, including the various portions of the examination for entry-level firefighters. The manual explains how to determine the required qualifications for candidates. An examination of the duties of a firefighter and a company officer are explored.

The next step introduced is a description of what a candidate should do prior to the interview and the importance of making a good first impression. A guide for the development of a resume is provided.

The following step is an examination of the interview process itself. Guidelines are provided as to how to answer general questions, how to field opening questions, and how to analyze and solve general problems, personnel problems, and firefighting problems.

The last chapter brings everything together. Questions are provided for both entry-level firefighter and company officer candidates. These questions are to be used to practice, practice, practice. The various methods of practicing are examined.

Features of This Book

Contained within these pages are also many helpful tips and tricks for learning and practicing your way to success in the oral interview:

- **Background information** on the examination and interview process sets you up for success, providing insight into interviewer expectations and advice on presenting yourself in a professional manner through applications and resumes.
- **Various example questions** are presented in the book for practice—opening questions, as well as questions on problem-solving, personnel problems, and firefighting situations—to ensure that you are prepared to answer these questions at the time of the interview.
- **Entrance and promotional level** interviews are explored, including examples of questions from each interview, to provide you with the information you need to succeed as a firefighter and as a proficient leader in the fire service.
- **Tips for success** are highlighted throughout the text to pinpoint specific advice for various situations that you may encounter during the interview process.
- **Interview practice** is emphasized throughout the book, including practical information on setting up and conducting practice sessions to efficiently prepare you for the actual interview.

Available Resources for This Book

If you are an instructor who wishes to use this as a training manual to complement an introduction to firefighting or fire officer course, an accompanying Instructor's Guide on CD-ROM is available, which includes the following:

- **Answers to review questions** that appear at the end of each chapter
- **PowerPoint presentations** to highlight the important points in each chapter

ABOUT THE AUTHOR

Gene Mahoney was released to inactive duty as a pilot from the U.S Navy in 1946. He served an additional sixteen years in the Reserve, retiring as a lieutenant commander. He flew both reciprocating engine and jet aircraft during his time with the Navy.

Gene joined the Los Angeles Fire Department as a firefighter in 1947. He retired as a battalion chief in 1969. During his time with the department, he was assigned to various areas of the city. His assignments included five years as battalion commander of the most active fire-fighting battalion in the city, five years in the harbor area, five years in the downtown area, and additional time in the high-rise area of the city and as battalion commander in charge of the firefighting forces at the Los Angeles International Airport. His special-duty assignments included several years in the training section. He was responsible for the public relations section of the department at the time of his retirement.

While with the Los Angeles Fire Department, Gene attended the University of Southern California, where he received his B.S. degree in Public Administration with a minor in Fire Administration in 1958. Three years later he earned his M.S. degree in Education.

Gene retired from the Los Angeles Fire Department to accept the position of fire chief for the city of Garden Grove, California. He was later advanced to the position of public safety director and then accepted the assignment as assistant city manager for public safety. In these positions, he was responsible for the operation of both the fire and police departments. He left the city of Garden Grove to accept the position of fire chief for the Arcadia, California Fire Department. He retired from this position in 1975.

Gene, together with another member of the Los Angeles Fire Department, was responsible for the development of the fire science curriculum at Los Angeles Harbor College in Wilmington, California. He served there as a part-time instructor for twelve years. He also taught fire administration courses for two years at Long Beach State College in Long Beach, California. Upon retiring as fire chief from the city of Arcadia, he accepted the position of fire science coordinator at Rio Hondo College in Whittier, California. While there, he developed the fire science curriculum into one of the most complete programs in the United States. The program included a Fire Academy that provided all the training required for certification as a Fire Fighter I in California. He retired from Rio Hondo College as a professor of fire science in 1988.

In addition to authoring several articles in professional magazines, Gene has authored several textbooks and study guides in the field of fire science, including: *Fire Department Hydraulics; Introduction to Fire Apparatus and Equipment; Fire Department Oral Interviews: Practices and Procedures;* and *Fire Suppression Practices and Procedures.* The study guides include: *Introduction to Fire Apparatus and Equipment; Firefighters Promotion Examinations;* and *Effective Supervisory Practices.* He also published a novel entitled *Anatomy of an Arsonist.*

During his career, Gene has been very active in professional and service organizations. He served as

District Chairman, Boy Scouts of America
President, United Way
District Chairman, Salvation Army
President, International Association of Toastmasters
President, Rio Hondo College Faculty Association

ACKNOWLEDGMENTS

This manual has been in the making for over forty years. The number of direct and indirect contributors runs into the hundreds, many of whose faces and names I can no longer recall. The list of indirect contributors includes a number of once starry-eyed youngsters who wanted to become firefighters. These youngsters came to me as a member of an organized class or as an individual, seeking help on how to best prepare for an oral interview, or perhaps after an interview with the question, "What did I do wrong?"

Most of the contributors, however, have been seasoned firefighters or ranking officers who sought the same information—and the same answers. Some of the contributors were those to whom I went early in my own career, seeking information on what I could do to better prepare myself for this challenge that seemed to have no set guidelines or parameters. Others were those who sat on practice or actual review boards with me, evaluating candidates and making suggestions for improvements. I cannot help but recognize the input the members of the San Pedro Toastmasters Club had on this project. Their constant evaluation of the presentations of members, together with their suggestions for improvement, certainly had an influence upon the ideas that went into this manual. All of these contacts, with their questions and suggestions, helped provide the foundation for the procedures and guidelines offered here.

There are some who have made a direct contribution to this particular book. The list includes

- Rick McClure of the Los Angeles Fire Department, who graciously furnished the fire-scene photographs
- Training Officer Jim Perkins and other members of the Joplin, Missouri Fire Department
- members of the Redings Mill Fire Department
- my grandson Ian Glassman, who posed as the entry-level firefighter candidate in photographs
- other members of the family who also gave their time as models for photographs: my stepdaughters Jeri Stapp and Cathi Glassman, my grandson Nathan Glassman, and my wife Ethel

A family friend, Ashley McWilliams, also gave her time during the photograph-taking portion of this project.

I also wish to thank the staff at Thomson Delmar Learning for their help in putting the manual together. Special thanks go to Acquisition Editor Alison Weintraub, who encouraged me to prepare the manual; Jennifer Thompson, Product Manager, who guided me through the preparation process; and Jennifer Hanley, Content Project Manager, who helped smooth out the rough spots in the final draft.

Special thanks is also extended to Interactive Composition Corporation (ICC) Project Manager Panchi Das and copy editor Catherine Albano, with whom I worked so closely in putting together the final steps in the production of the book.

Finally, I thank the following individuals who generously gave their time and shared their experience in reviewing the manuscript. Their recommendations were very beneficial in the preparation of the final draft.

Anthony Carlorel
Senior Instructor
Burlington County Emergency Services Training Center
West Hampton, NJ

Alan Joos
Assistant Director-Certification
Utah Fire and Rescue Academy
Provo, UT

Paul Madden
Fire Science Program Director
Keiser College
Sarasota, FL

Rufus T. Summers
Fire Program Director
Houston Community College
Houston, TX

Noah James West
Fire and Emergency Services Coordinator
Louisiana State University
Eunice, LA

THE EXAMINATION PROCESS

LEARNING OBJECTIVES

Upon completing this chapter, the reader should be able to:

- Explain the overall examination process.
- List the important parts that may be included in an examination process.
- Explain how the final grades are determined on a weighted examination.
- Explain how appointments are made under the Rule of Three, the Rule of One, the Open Rule, and the Tier System.
- Explain the difference between the status of a member on probation and one who has tenure.

INTRODUCTION

The objective of this manual is to assist those desiring to become firefighters and those seeking promotion within the fire service in the preparation for the oral interview portion of the examination. In preparing for an oral interview, it is essential that candidates be familiar with the relationship of this portion of the testing process to the overall result. While the examination process itself may vary from one jurisdiction to another, the information contained in this chapter should provide a candidate with sufficient insight to relate to the practices within his or her own community or to any community to which he or she may wish to file an application.

There is one thought that entry-level firefighter candidates should keep in mind. While the examination process may seem demanding, it is a two-way street. Candidates are very anxious to get hired, and the hiring community is very anxious to assist the candidates so that they can hire the best potential firefighters available.

THE CIVIL SERVICE SYSTEM

Many of the positions in government during the nineteenth century were obtained through the practice of "to the victor belong the spoils." The **spoils system,** as it was known, was based upon the concept that public positions are considered gifts of the controlling authority to be used for either personal or political purposes, or both. Jobs under this system were normally dealt out first to relatives, then to friends, and third to political associates, normally those who had helped elected officials to office. Many government jobs were available under this system if a job seeker knew the right official and had sufficient funds to fill the coffer. In many cases positions were occupied by individuals who had neither the training nor the knowledge to carry out the functions of the job adequately. Government under this system was, without doubt, an organization run by amateurs.

During the 1870s and 1880s, a strong movement for some type of reform took place. The reform momentum culminated in 1883 with the passage of the **Civil Service Act.** The concept behind its development was to create a government in which those best qualified are hired rather than to continue appointing in accordance with the spoils system. Although the act was the strong point in the movement to change government from services run by amateurs to those run by professionals, it took many years for the reform to penetrate all levels of government. Even today, many of the top positions at various levels of government are still filled by the spoils concept. With all due respect, however, most appointing authorities make every effort to fill these spoil positions with well-qualified personnel. With the exception of many high-level spoils appointments, most positions in government today are filled in accordance with the civil service reform philosophy—the best-qualified person should be appointed to the position. This concept is known as the **merit system.**

There are some people who will argue that the concept of the merit system is being eroded through the Civil Rights Act and court-mandated minority hiring quotas. Regardless, those hired are still subject to the examination process. Individuals are not normally hired unless they meet the minimum standards established by the examining jurisdiction. Consequently, in order to compete it is essential that candidates for governmental positions be thoroughly familiar with the total examination process. It is much like a game. To compete, you must know the rules.

THE HIRING PROCESS

The hiring process is relatively simple. A position in government becomes vacant, or it is anticipated that one or more positions will become vacant. An announcement is made, usually in bulletin form, that an examination for the position will be conducted.

Those interested file an application and compete in the examination process. An **Eligible List** based on the results of the examination is established. The Eligible List indicates the names of candidates who have successfully passed and are eligible for hiring or promotion. The Eligible List may also be referred to as the Eligibility List or as a List of Eligibles. However, for simplification purposes, it will be referred to in this manual as the Eligible List.

Candidates are hired from the Eligible List in accordance with the civil service rules and regulations. Those hired must then serve a **probationary period** before the appointment becomes permanent. The probationary period is referred to as the "working test period."

THE EXAMINATION ANNOUNCEMENT

To understand the hiring process, it is first necessary that all candidates for governmental positions realize that examinations for positions within a particular department in a large city are not normally conducted by that department. However, in smaller communities members of the fire department become more actively involved. In the larger cities, the fire department does not normally conduct the examination for entry-level firefighters, nor does the water department conduct the examinations for secretaries or file clerks who will work within that department. The Civil Service Department, the Personnel Department, or the Human Resources Department, depending upon the organization of the governmental agency, generally conducts examinations. Professionally trained examiners who conduct the examinations and establish the Eligible List staff these departments. The Eligible List will be discussed in more detail later in this chapter.

The examining department is usually under the control of a Civil Service Commission, a Personnel Board, or a Human Resources Department. These bodies are responsible for seeing that the principles of the merit system are enforced. They establish the rules and regulations for examinations and examination procedures. The process must be conducted in accordance with their regulations. Any candidate who believes that he or she has been unfairly treated, or that the rules and regulations have been violated, has the right to file an appeal or protest with these agencies for relief. For purposes of simplification, both the Civil Service Department, the Personnel Department, and the Human Relations Department will be referred to in the remainder of this manual as the Civil Service Department.

Examinations are conducted by governmental agencies whenever the need arises. In larger jurisdictions, the Civil Service Department generally attempts to maintain an Eligible List at all times. When this is done, examinations for a particular position are given on a fairly regular schedule, as an example, every two years. Of course, if the Eligible List is exhausted sooner than the scheduled period, then an examination is held. In a few jurisdictions, examinations are processed on a continuous basis for some positions. Smaller jurisdictions, however, operate differently. Examinations may not be given until there is a vacancy to be filled.

It is extremely difficult for an individual desiring a particular position in government to know when an examination for that position will be given by various jurisdictions. It is necessary for candidates to keep in touch with Civil Service Departments in order to stay abreast of examination scheduling. As an aid to potential employees, some jurisdictions will accept interest cards. Candidates can complete an interest card and leave it with the Civil Service Department. When an examination will be given, the Civil Service Department fills in the necessary information and mails the card to the candidate. An example of an interest card used by the fictitious city of Hillstown is shown in Figure 1-1.

While a number of communities issue their own interest cards, most that do will also accept cards made and submitted by candidates. A sample of an interest card that might

Place your name and address on the other side of this card together with proper postage. You will be notified if an examination is announced within one year from the date you file this card.

Phone ______________________ Date ______________________

CLASSIFICATION ______________________________________

The filing date for the above classification has been announced. A copy of the announcement listing the filing requirements is attached to this card. If you are still interested in applying, you must file an application with the Human Resources Office, Hillstown City Hall, 257 N. Main Street, Hillstown, Anystate, 00000, prior to the closing date.

Figure 1-1 An example of an interest card issued by the city.

City of ______________________________________

Applications for the position of firefighter will be

accepted commencing ____________________________

The last date to file is ____________________________

Figure 1-2 A sample interest card made by a candidate.

be made by a candidate is shown in Figure 1-2. It can be an ordinary postcard obtained from the post office or a card designed by an applicant. On the opposite side of the card, the candidate should place his or her mailing address and adequate postage. Candidates should ensure that the card designed meets post office regulations for mailing.

A number of communities will place an interested candidate on an interest list by contacting the community's Web site on the Internet. The Web site will normally also provide a list of the requirements for filing for the position. This is an effective method for a candidate to stay in contact with a community for whom he or she would like to work (see Figure 1-3).

Civil Service Departments use various methods to announce examinations. Some run advertisements in local newspapers (see Figure 1-4). Others post the examination bulletin in various public places, such as libraries. Almost all jurisdictions post the announcement on a bulletin board in the Civil Service Department itself. More and more jurisdictions are also making use of the Internet for announcing job openings.

Most jurisdictions allow at least two weeks from the initial posting of the examination announcement to the last date for filing an application. It is important for candidates to

The Hillstown Fire Department just ended its 2005 open application process. The next estimated time line for accepting job applications will be in the year 2007. However, the department accepts interest card submissions all year round. Please enter the following information and submit. When your submission is received your name is entered into the fire department's recruitment database. Once open application begins, you will be notified by mail. ONLY ONE SUBMISSION PER PERSON. Please check out the Hillstown Fire Department recruitment page under the Professional Development Bureau for further details and other requirements. The following are the minimum qualifications to apply for the position of firefighter. AT THE TIME OF APPLICATION, a candidate must be between 18 and 37 years of age; be a resident of Hillstown; possess a valid driver's license; and have a high school diploma or equivalent. Thank you for your interest in the Hillstown Fire Department.

***Asterisks denote required fields.**

Name*

Address*

Phone Number

E-Mail

SUBMIT

Figure 1-3 An example of an Internet interest card.

obtain or review the announcement bulletin, as it contains information that is both useful in understanding the rules of the game and very important in assisting a candidate in preparing for the oral interview.

The Examination Bulletin

Figure 1-5 is a sample examination bulletin for the position of firefighter. An announcement for the position of captain, battalion chief, and so forth would be similar. There are several important items in the bulletin that should be noted.

FIREFIGHTER/EMT TESTING
Monday, June 6, 2005

Physical Agility Test: 9:00 a.m. at the Public Safety and Justice Center (Former City Hall)
Located at 303 East Third Street, Hillstown Station #1

Written Test: 1:00 p.m. at City Hall
Located at 257 North Main Street, Hillstown
Basement Conference Room

Minimum criteria for applicants to **participate** in department testing:
(1) Minimum 18 years of age.
(2) No prior criminal convictions.
(3) Possession of valid driver's license with good driving record.
(4) Possession of Firefighter II certification from this state; OR possession of Firefighter II certification from another state and eligibility to obtain Firefighter II certification prior to employment; OR proof of current enrollment in a program leading to Firefighter II certification.
(5) Possession of current EMT/B license from this state; OR possession of EMT/B licensure from another state or National Registry of Emergency Medical Technicians (NREMT) certification and eligibility to obtain EMT/B licensure prior to employment; OR proof of current enrollment in a program leading to EMT/B licensure/certification.

Preference will be given to applicants who possess Paramedic licensure from this state or National Registry of Emergency Medical Technicians Paramedic certification.

Based on test scores and a preliminary background investigation, applicants will be placed on a "pre-qualified" list for one year for hiring purposes. Candidates must meet additional requirement prior to hire, including possession of Firefighter II certification and EMT/B licensure. Proof of Firefighter II and EMT/B certification or course enrollment must be submitted at time of application. Applicants are expected to be present at the specified place and time unless otherwise notified.

To participate in testing, application must be received in the Human Resources Office, 257 North Main Street, Hillstown, Anystate, 00000, **by Wednesday June 1, 2005 at 5:00 p.m.** Visit our Web site at www.hillstownanystate.org and click on "employment" to download an application to mail in. Pre-employment physical and drug testing are required of all successful applicants, in addition to a background check. Starting annual salary $25,373 as earned. The City of Hillstown is an affirmative Action, ADA, and Equal Opportunity Employer.
M/F/V/D.

Figure 1-4 A sample newspaper announcement of an examination.

CITY OF HILLSTOWN
Job Opportunity
FIREFIGHTER

DUTIES: Under direct supervision, responds to fire alarms and other requests for emergency and non-emergency assistance; connects and lays hose lines; enters burning buildings with hose lines; assists in directing streams of water on fires; raises and lowers ladders; makes forcible entry into buildings; participates in fire prevention inspections, inspecting structures, and informing public on fire hazards; participates in training activities; maintains station, apparatus, and equipment; performs other related work as required.

MINIMUM REQUIREMENTS:

1. Age: At least 18 and not more than 37 at time of appointment.
2. Education: High school graduation or GED equivalent.
3. Possession of a valid driver's license.
4. EMT-1 certification
5. Tobacco use is restricted on and off duty.
6. No felony convictions.
7. No Class A misdemeanor convictions within 24 months of application.

DESIRABLE QUALIFICATIONS

1. Bilingual abilities in Spanish and English.
2. Firefighter I and II certification.
3. Experience in a recognized trade requiring the use of hand and power tools.
4. Skills in computer operation.

SELECTION PROCESS

Written examination (weight, 40%): This examination may include material designed to measure mechanical comprehension, spatial orientation, situation reasoning, understanding of illustrated materials, basic arithmetic, and reading and writing abilities.

Physical ability (weight, 30%): This examination will consist of a series of tests to determine the applicant's ability to perform critical physical tasks of the firefighter's job. The examination will consist of the following five unskilled tasks:

1. Dragging a hose and nozzle over a measured course.
2. Connecting hose couplings to a hydrant.
3. Climbing a 75 ft. ladder to the top of a five-story building and descending through the interior stairway.
4. Loading and unloading sections of hose (approximately 60 lbs.) onto a fire apparatus hosebed.
5. Pulling a rolled section of hose (approximately 60 lbs.) to the top of a hose tower four times.

Oral interview (weight, 30%): The oral interview is designed to evaluate a candidate's experience, training, and personal qualifications.

BACKGROUND INVESTIGATION: Prior to employment, a thorough investigation will be conducted into the background of eligible applicants.

PSYCHOLOGICAL TEST: Prior to appointment, all eligible applicants will be subjected to a series of psychological tests and interviews.

LAST DATE TO FILE: Closing date for filing applications is 5 p.m. on June 1. Applications are to be filed with the Human Resources Office, Hillstown City Hall, 257 North Main Street.

AN EQUAL OPPORTUNITY EMPLOYER

Figure 1-5 A sample examination bulletin.

Minimum Qualifications

A candidate should ensure that he or she meets the minimum qualifications before filing an application. The minimum qualifications will vary from community to community. One of the minimum requirements may be that the candidate has a valid driver's license. More and more communities require that candidates for entry-level firefighters have a current Emergency Medical Technician certification or be enrolled in an EMT certification program. Some require that a candidate be Firefighter I or Firefighter II certified from an accredited entity (see Figure 1-4).

Note that the age requirement may be a problem. There may be a minimum and maximum age requirement. The minimum age requirement will generally vary between 18 and 21. If established, the maximum may be as low as 30 or as high as 37.

There are some people who believe that the maximum age requirement for firefighter candidates violates the age discrimination act passed by Congress. This is not true.

The age discrimination act was passed by Congress in 1967 to protect individuals between the age of 40 and 65 from discrimination in employment. In 1986, Congress amended the age discrimination act. This amendment exempted, through December 1993, state and local governments when hiring or retiring firefighters or law enforcement officers from limitations, provided those limitations were in effect in March 1983. In 1996, Congress passed another age discrimination employment amendment, which permanently reinstated an exemption that permits state and local governments to use age as a basis for hiring and retiring firefighters and law enforcement officers. This amendment is still in effect.

It may be necessary to meet the age requirement on the last date to file, the date of the written test, or perhaps the date of appointment. On the bulletin shown in Figure 1-5, the age requirement is a minimum of 18 and a maximum of 37 at the time the applicant is appointed to the position.

Last Date to File

Filing means submitting an application to take the examination. The last date to file shown on an examination is generally firm. For example, an application for the examination shown in Figure 1-5 would not be accepted after 5:00 p.m. on June 1. However, some jurisdictions do not establish a time factor and will accept applications postmarked by the last date to file. Regardless of the requirements established, a candidate should not let sickness, a freeway tie-up, an untimely flat tire, or some other unexpected incident eliminate him or her from competing in the examination process.

There is, however, another important reason for filing applications early. Some jurisdictions use the date and time of filing to break ties on examinations. If two candidates receive identical scores on the examination, the one that filed his or her application first is usually placed higher on the Eligible List. As an example, candidates number 4 and 5 on the Eligible List shown in Figure 1-10 received identical scores of 95.13. Number 4 was placed higher on the list because his application had been filed prior to that of number 5. For promotional examinations, a number of jurisdictions use seniority rather than filing dates to break ties.

All jurisdictions do not use the filing date to break ties on entrance examinations. Residence may be a factor in some communities, ties may be broken by drawing a name of one of the tied candidates out of a hat in other communities, or some other method may be used in accordance with the jurisdiction's civil service rules and regulations.

Date of Examination

The date of the written examination and perhaps the physical ability or practical portion of the examination are given on some bulletins. Others do not indicate the dates. Candidates who file applications are later advised by mail as to the dates, times, and locations of various phases of the examination. A candidate should place the given dates on his or her appointment calendar as a constant reminder.

Duties

The list of duties given in the bulletin provide a candidate with some understanding of the types of activities with which he or she will be involved if hired. Candidates should become thoroughly familiar with these duties. Knowledge of this information is essential for the oral interview portion of the examination process. This point will be discussed in more detail in a later chapter.

Desirable Qualifications

It should be noted that there are minimum qualifications required in order to compete in the selection process; however, there may be other skills and abilities the appointing authority is seeking in the person he or she hopes to hire. Many times these desirable qualifications are listed on the announcement bulletin (see Figure 1-5). While these additional qualifications are not essential in order to be hired, all other things being equal, a candidate having all or a portion of them would certainly have an advantage in the examination process. The advantage normally manifests itself in the oral interview. It therefore behooves a candidate having one or more of the desirable qualifications to identify them on the application and the resume, if a resume is submitted.

Selection Process

The selection process outlines several phases of the examination procedure. Other phases are identified elsewhere in the announcement bulletin. The number of parts in the examination procedure will vary from one jurisdiction to another and may vary from examination to examination within the same community. While the examination process for company and higher officers may be limited to the written examination and an oral interview or assessment center, an entrance examination may vary from a small to a large number. Some or all of the following may be included in the examination process.

- written test
- oral interview
- physical ability test
- psychological test
- medical examination

- background check
- drug testing
- polygraph

Written Test A written test is almost a standard part of every entry- and promotional-level examination for fire departments. A written test is designed to be both valid and reliable. A test is valid when it measures what it is supposed to measure. A test is reliable when it can be given to an individual today and given to the same individual at a later date with the candidate receiving approximately the same score.

Fire department written tests for promotional candidates are designed to test a candidate's knowledge of the technical requirements for the position. As an example, a written examination for the position of battalion chief might include questions on organization, administration, budgets, personnel management, firefighting, training, arson, department regulations, and federal, state, and local laws. Examinations for company-level officers normally include much of the same subject material as that for battalion chief, except the questions are presented at a lower level. For example, supervision questions are generally used rather than administration questions and firefighting questions involve company operations rather than large-scale operations.

Written examinations for entry-level positions, however, would be designed to test general knowledge applicable to the position. Candidates would be expected to understand written material, to be able to spell, and to be capable of using common sense to solve situational problems. The examination might be designed to test for mechanical comprehension, spatial ability, basic arithmetic, and understanding of illustrated materials.

It is extremely difficult for candidates to compete on a written examination without adequate preparation. It is important to become familiar with the types of questions used in civil service examinations and to acquire a thorough understanding of the scope of the examination. The best method of acquiring an understanding of the scope is to obtain a previous examination for the position that was given by the testing agency. This, however, is not always possible. The next best thing is to review typical questions for the position. Training manuals for this purpose may be purchased in technical bookstores or by direct mail order. The Internet provides a good source for available books. Furthermore, copies of training manuals may be available in the reference section of many libraries. There are also a number of schools available for assisting candidates in preparing for the examination.

Oral Interview Regardless of how well a candidate does on other portions of the examination process, it is generally the oral interview or assessment center that determines whether or not he or she is hired. In some communities, the other phases of the examination process are graded on a pass-fail basis with the Eligible List established solely on the basis of grades on the oral interview. The objective of this manual is to assist candidates in preparing for this critical portion of the examination process.

Physical Ability Test Some communities refer to this portion of the examination process as a physical agility test (see Figure 1-4) while others refer to it as a physical ability test. It is also referred to as a physical fitness test (see Figure 1-8). From a technical standpoint, there is a difference between a physical agility test, a physical ability test, and a physical fitness test. However, from a practical point, the test determines the same factors regardless of whether it is referred to as a physical agility test, a physical ability test, a physical fitness test, or perhaps some other designation. Consequently, for gaining an understanding of the objective of the test, the terms physical agility, physical ability, and physical fitness should be considered interchangeable.

The physical ability test is part of the practical portion of the examination process. In most cases the practical portion of the examination process will test a candidate's ability to perform the duties of the position. For example, an apparatus operator might have to demonstrate the ability to drive an apparatus, pump from draft, pump from a hydrant, operate an aerial ladder, or operate an aerial platform. An examination for the position of entry-level firefighter, however, does not test a candidate's ability to actually perform the duties because candidates in many communities are neither expected nor normally required to have had previous training in these areas. Consequently, activities that are job-related and test a candidate's strength, agility, endurance, and coordination needed to perform the job are normally given (see Figures 1-6 and 1-7).

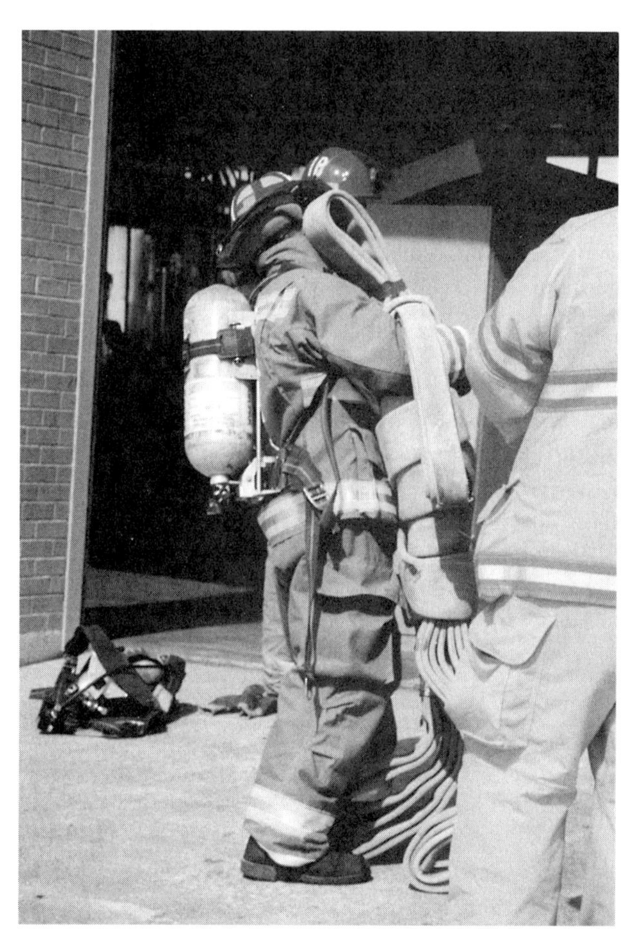

Figure 1-6 Simulated rescue portion of the physical agility test.

Figure 1-7 The hydrant and hose evolution portion of the physical agility test.

There is no national standard physical ability test that is used by all departments. This phase of the examination process will vary from community to community. Some communities refer to the physical ability examination as a physical fitness evaluation test. For example, a job-related physical fitness evaluation test for a firefighter given by the Joplin Fire Department in Joplin, Missouri is shown in Figure 1-8.

While there is no national standard that is used by all departments for assessing the physical ability of entry-level candidates, there is one that is available for use by all career departments. The standard is quite extensive and physically demanding. Many career departments that would like to use it do not do so due to fiscal restrains, as the equipment required for establishing the test is expensive. This standard is referred to as the CPAT.

CPAT stands for **Candidate Physical Ability Test.** Both the International Association of Firefighters (IAFF) and the International Association of Fire Chiefs (IAFC) assisted in the development of the test and actively support it. The test includes eight sequential events:

1. Stair Climb
2. Hose Drag
3. Equipment Carry
4. Ladder Raise and Extension
5. Forcible Entry
6. Search
7. Rescue
8. Ceiling Breach and Pull

PYSICAL FITNESS EVALUATION
JOPLIN FIRE DEPARTMENT

There are a total of eight (8) exercises. All exercises will be completed on the same day. All exercises are scored on a point system basis. The time each applicant takes to complete an exercise will be evaluated and a point given. The point value is determined based on a predetermined score as set by our test provider and an average score set by averaging the time of firefighters currently employed. The physical fitness evaluation is only one of four factors to be evaluated on each applicant. THIS IS NOT A PASS/FAIL TEST. Candidates will be allowed to retake only one of the eight tests if they fail to complete the test on the first try.

1. ABOVE GROUND FIRE ATTACK SIMULATION—Carry a 75-pound bundle of fire hose down a flight of stairs, all the way back up, halfway down the stairs, and back up to the top.
2. HYDRANT AND HOSE EVOLUTION—Pull three sections of hose, which weigh approximately 50 pounds each, a distance of 100 feet and couple them with a fixed connection.
3. LADDER EVOLUTION—Remove a 14-foot ladder from its mounting on the side of a fire truck, position it against a wall, climb the ladder, return it to the ground, and remount the ladder on the truck.
4. SIMULATED RESCUE—Drag a weight, simulating a person weighing 150 pounds, a total of 140 feet on a level surface.
5. LADDER CLIMB STATION—Climb an extension ladder approximately 20 feet and dismount onto roof. Remount ladder and return to ground.
6. ELEVATED LINE ADVANCEMENT SIMULATION—Standing on a roof, pull 50 feet of rope, then 50 feet of 2-1/2-inch hose that is tied to the line, over a pulley assembly until the hose nozzle reaches an indicated location.
7. JOIST WALK SIMULATION—Carrying a 50-pound coil of hose, walk the length of a 28-foot ladder laid on the ground, stepping only on the side rails and without stepping off of the ladder. Stepping off of the ladder or stepping on the rungs will be considered an incomplete exercise.
8. FORCIBLE ENTRY SIMULATION—Swing a sledge hammer (approximate weight of a fire axe) 15 times at an 18- to 24-inch-high target surface, using a full overhead two-arm swing.

Figure 1-8 An example of a physical fitness evaluation test.

If a candidate is informed that this test will be used as a portion of the examination process, he or she can find information on the testing procedure on the Internet. It is advised that this source be checked out.

Some communities are including a practical phase to the examination process for fire department officer candidates. The practical phase is generally presented in the form of an assessment center. Some jurisdictions restrict the use of assessment centers to the selection process for top management positions. Others are using the concept for company-level positions as well. The assessment center concept has also been used at the entry level by a few jurisdictions. Normally, when assessment centers are utilized, the oral interview phase of the examination process is included within the testing procedure in a modified form, or at the minimum, candidates are required to demonstrate their ability to present themselves before a group in either an impromptu or a prepared state.

It is interesting to note that some communities are subjecting promotional candidates to a physical ability test. The physical ability test might be the identical test required for an entry-level position or that test modified to a minor degree.

Psychological Examination Some individuals can handle stress very well, and others cannot. Firefighters are exposed to a considerable amount of emotional and mental stress at fire and other emergency situations. The capability to withstand stressful conditions is essential for efficient performance. A firefighter who fails to perform satisfactorily in stressful situations may place not only himself or herself in danger, but may also contribute to a considerable loss of life, property, or both. Consequently, some jurisdictions are subjecting firefighting candidates to psychological testing in order to determine their emotional and mental characteristics.

The psychological test may include both a written test and an interview with a psychologist. The psychologist may ask about the candidate's interests, attitudes, background, hobbies, and his or her relationships with friends and family members. A candidate should be open and honest in answering these questions. This is more important than the answers to the questions themselves.

This portion of the examination process is usually a pass-fail situation with the results having no bearing on a candidate's position on the Eligible List. A candidate failing this part of the examination process will not, however, normally be hired regardless of how well he or she did on other portions of the examination process.

Because of the cost involved, a candidate may not be subjected to the psychological testing phase unless he or she has passed all the weighted portions of the examination, is on the Eligible List, and is being considered for immediate hiring. Weighted refers to those portions of the examination process that have an effect on the final score. Unfortunately, there is no way for a candidate to prepare for this part of the examination process.

Medical Examination The examination procedure for most fire department entry-level positions requires the candidate to pass a medical examination satisfactorily prior to being hired. This examination is similar to a standard physical examination given by a candidate's family doctor. A candidate's weight, height, blood pressure, temperature, lungs, heart, eyes, ears, nose, and mouth will normally be checked. A blood test and urine test will also most likely be given (see Figure 1-9).

Some departments also require candidates for promotional positions to pass a medical examination prior to being promoted. Standards vary from jurisdiction to jurisdiction; however, most fire departments have high standards with back and cardiac problems usually the primary cause for rejection. Like the psychological part of the examination process, the medical examination may not be given until a candidate is being seriously considered for hiring. It is usually rated on a pass-fail basis. A candidate failing the medical examination will not be hired.

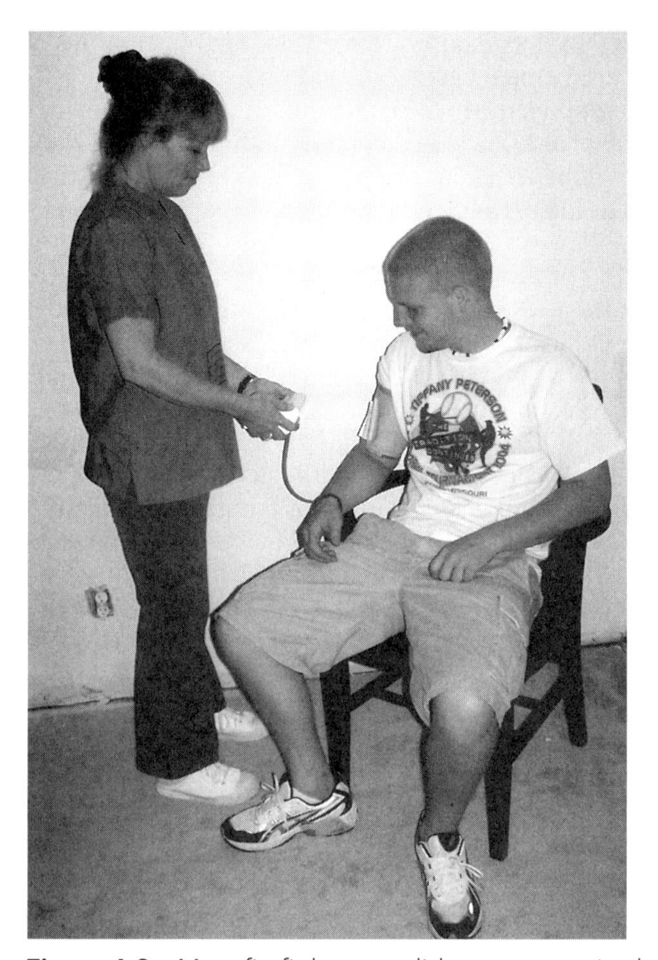

Figure 1-9 Most firefighter candidates are required to pass a medical examination prior to being hired.

Background Check A fire department is the only organization in the world that has the legal authority to keep an owner of a home or place of business out of the structure while the department makes a thorough inspection of all parts of the occupancy. Because of this, any individual who has this freedom should be both honest and have high integrity.

Not only should a firefighter have these characteristics, but it is also important that he or she have the ability to get along with people and have a good reputation within the community. These factors cannot be determined on a written examination or completely on an oral or psychological examination. Background checks are given to ensure that entry-level firefighters have the characteristics desired by the department.

Background checks may be conducted by fire department personnel on large departments or by law enforcement personnel on smaller departments. Some smaller departments use an outside agency for this purpose. Such checks may include fingerprinting and polygraph examinations as well as reports from educational institutions, former employers, acquaintances, law enforcement agencies, and other sources.

In addition to written reports, the investigation may include talking to people within a community who have personal knowledge about an individual. Typically, all references listed on the application will be checked. The person making the investigation will talk to personal friends of the candidate, teachers, former employers, and others, such as church pastors.

In addition to gathering information from individuals such as these, the investigator will check public records for arrests, traffic violations, and other criminal activity. Results of the total investigation will usually be given to a civil service committee or the appointing authority. The results of the investigation will remain confidential and, if unsatisfactory, may be the basis for rejecting the candidate for hiring.

Drug Testing Many jurisdictions check candidates for the use of drugs prior to hiring. The drug check may be included as a portion of the medical examination. However, some jurisdictions check a candidate without any previous notification. Random testing should be part of the notification hiring process so that the candidate is aware it could happen. Candidates who use drugs or have used them within a specified period are normally eliminated from consideration for hiring.

Polygraph A polygraph test may be given to a candidate individually or as part of the background check. Polygraph tests are generally referred to as lie detector tests. However, they are not capable of determining whether or not an individual is lying. They are only capable of detecting when deceptive behavior is being displayed. The value of the results of a polygraph test to those conducting a background check is that the test may provide an investigator with an area that should be further investigated.

ESTABLISHING AN ELIGIBLE LIST

As previously noted, some portions of the examination process are merely pass-fail, while other portions contribute to a candidate's final grade. The written examination, the oral interview, and the practical examination may be graded with the individual scores contributing to a candidate's final standing. This is referred to as a weighted examination. However, a number of departments establish the Eligible List solely from the oral interview or practical examination with all other parts of the examination process graded on a pass-fail basis. For the purpose of understanding the procedure involved when various parts of the examination process are weighted, the following example is offered.

	Grade		Weight
Written examination	87.34	×	.40 = 34.936
Oral interview	92.46	×	.30 = 27.738
Practical examination	79.39	×	.30 = 23.817
Final grade			= 86.491

Final Grade

Using the example previously given, the written examination is weighted 40 percent, the oral interview 30 percent, and the practical examination 30 percent. The candidate received a grade of 87.34 on the written examination, a 92.46 on the oral interview, and a 79.39 on the practical examination. The grade on an examination is multiplied by the weight given on the examination and the result on each portion of the examination process is added to obtain the candidate's final grade, provided that additional points are not given for military credit, seniority, or other factors. For example, if this examination had been used for entry-level firefighters, and 10 points were given for veteran's credit, then the 10 points would be added to the final grade (86.492 + 10.00 = 96.491), provided that the candidate qualifies as a veteran. The candidate would be given a grade of 96.491. This grade would be used to determine his or her placement on the Eligible List.

Some jurisdictions give credit on promotional examinations for seniority. Suppose this examination was a promotional exam with a weight of 0.25 given for each year of service. If a candidate has nine years in the department, then he or she would receive 2.25 points for seniority (9 × 0.25). This would be added to the final grade of 86.491 obtained in the examination process (86.491 + 2.25 = 88.741) and the candidate would receive a final grade of 88.741. This grade would be used to determine his or her placement on the Eligible List.

Some jurisdictions give credit on promotional examinations for an evaluation of a candidate's past performance. The final grade on the examination is determined in the same manner as that for seniority. This credit is often referred to as merit credit.

Some larger communities have completely departed from a final score based upon weighted parts of the examination process. Many establish the Eligible List from grades received from the oral interview or written examination only. Those that have established the Eligible List solely on the grade on the written test or on the oral interview consider the other parts of the examination process on a pass-fail basis.

The Eligible List

Once the final grades are determined, an Eligible List is established. A sample Eligible List is shown in Figure 1-10. The Eligible List is a list of those candidates from which appointments can be made. Every candidate who passes the examination with a score at or above the established minimum is placed on the list. Generally, the minimum passing mark is 70 percent; however, it may be higher or lower in some jurisdictions.

Once the list is established, the normal procedure is to submit it to the Civil Service Commission for adoption. When adopted, the list is dated and remains in effect in accordance with the time span established in the civil service rules and regulations. Many lists remain in effect for two years. Others are good for a year, while some are in effect for one year and may be extended for an additional year upon the request of the appointing authority. Of course, there are probably many additional variations. Nevertheless, it is important to understand that appointments can be made only from an Eligible List during the time the list is in effect. Once the time limit runs out, a new examination must be conducted and a new list adopted before anyone can be hired.

HILLSTOWN FIREFIGHTER ELIGIBLE LIST			
1. Jose Ruiz	97.48	13. Mathew Duffy	83.74
2. William Riley	97.02	14. Willie Smith	81.23
3. Maria Gonzales	96.49	15. Michael White	79.36
4. Charles Johnson	95.13	16. James Carlson	79.35
5. Curt Keller	95.13	17. Walter Metcalf	78.97
6. Nathan Potter	94.27	18. Carl Garvey	77.14
7. Glen Carter	93.54	19. Harry Turner	77.02
8. Gary Miyata	91.97	20. Chris Potter	76.58
9. Debra Young	89.39	21. Scott Carroll	74.38
10. John Anderson	88.11	22. Lewis Adams	72.68
11. Ronald Dulley	87.48	23. Eldon Williamson	71.45
12. Jose Martinez	85.33	24. Donald Roberts	70.02

Figure 1-10 A sample eligible list.

Some jurisdictions provide provisions in their civil service rules for withdrawing a list prior to the time limit expiring. The circumstances that allow the list to be withdrawn are outlined in the rules.

THE APPOINTMENT PROCEDURE

The method by which candidates may be hired from an Eligible List is generally outlined in the civil service rules and regulations. Although there are undoubtedly many variations from the following, the procedure outlined will provide some understanding of the methods used.

The **appointing authority** is the person with the legal authority to hire and fire. He or she is normally the general manager of the department in which a vacancy occurs. For instance, the Fire Chief is the appointing authority for the fire department. In some departments, the general manager and appointing authority is referred to as the Fire Commissioner.

Whenever an opening occurs, he or she requests **certification** from the Civil Service Department. The Civil Service Department sends the appointing authority the names of the people who are eligible to be hired. Certification means that the Civil Service Department is certifying to the appointing authority that the candidates whose names are submitted have been examined in accordance with existing rules and have met the minimum standards for hiring.

The number of names submitted varies in accordance with the procedure used by the jurisdiction. The four principal methods in general use are the Rule of Three, the Rule of One, the Open Rule, and the Tier System.

Rule of Three

Whenever the Rule of Three is used, names are placed on the Eligible List according to their final grade in the examination process (see Figure 1-10).

Whenever certification for a single position is requested under the Rule of Three, the top three names from the Eligible List are given to the appointing authority. The appointing authority interviews the three and may hire the one he or she believes is best qualified. The appointing authority does not have to justify the selection. The names of the other two are returned to the list and are eligible for certification in the event another vacancy occurs. Some jurisdictions, however, allow candidates to be certified only three times from the same list.

For example, a candidate may be number three on the Eligible List when a single opening occurs. The names of numbers 1, 2, and 3 are submitted to the appointing authority for consideration. Number 1 is chosen and numbers 2 and 3 are returned to the Eligible List. Original number 3 is now number 2 on the list.

When the next opening occurs, numbers 1, 2, and 3 from the list are certified. Number 1 is hired and numbers 2 (originally number 3) and 3 are returned to the Eligible List. Original number 3 has now been certified twice and is now number 1 on the Eligible List. The next time an opening occurs, he or she will be certified the third time. If not hired, his or her name will not be returned to the Eligible List and the individual will no longer be considered for hiring.

If the appointing authority is not satisfied with any of the three candidates, he or she may reject all of them. However, a written report must be submitted to the Civil Service Commission giving the reason why each candidate was rejected. The next three candidates on the list will then be certified if the Civil Service Commission approves the rejections.

If more than one vacancy exists, then two candidates more than the number of vacancies are certified. For example, if there are five openings for captain, the names of the first seven from the Eligible List would be submitted to the appointing authority. The appointing authority would hire the five determined to be best qualified.

Although many communities employ the Rule of Three, the actual selection of one of the three candidates is generally used only for high-level positions or in small communities where only a limited number of people will be hired from a list. In larger jurisdictions where many people will be hired from a single list, appointments are usually made straight from the list in accordance with candidates' placement on the list.

Rule of One

An Eligible List similar to that used in jurisdictions employing the Rule of Three is used when the Rule of One prevails. Very few jurisdictions, however, have adopted this rule. The procedure for appointment is similar to that for the Rule of Three, except that only the name of the top candidate from the Eligible List is submitted to the appointing authority whenever a single opening occurs. The appointing authority must hire the candidate or justify by a written report to the Civil Service Commission why the candidate was rejected.

Open Rule

The Open Rule is based upon the concept that every candidate on the Eligible List has met the minimum qualifications and is therefore qualified to be hired. When a true Open Rule is used, the appointing authority may hire any candidate from the list whenever a vacancy exists. Many times the names on the list are arranged in alphabetical order with no indication as to each candidate's final grade on the examination process.

It is easy to see that selection under this rule would be extremely difficult if names are arranged in alphabetical order with no indication as to each candidate's final grade and there are only one or two vacancies. It is also apparent that this system could be abused and is somewhat of a departure from the philosophy of hiring "the best-qualified person" for the position. It does, however, allow jurisdictions subjected to court-mandated minority hiring or social pressure more leeway in the selection process.

Fortunately, only a few jurisdictions use a complete Open Rule system. There is a trend, however, toward using a modified system. For sake of a better term, this modified system might be referred to as a Tier System.

Tier System

When a Tier System is employed, names on the Eligible List are generally arranged in accordance with final scores on the examination process, but are further divided into tiers. Several methods are used for establishing the tiers. The total list may, for example, be divided into thirds, or it may be divided by grades. Using the sample Eligible List in Figure 1-10 as an example, if divided into thirds, Tier A would include the first eight candidates, Tier B the second eight, and Tier C the third eight. Or it could be divided so that those candidates with grades of 90.00 or above would be placed in Tier A (numbers 1 through 8); those with grades of 80.00 and up to, but not including 90.00 (numbers 9 through 14) placed in Tier B; and those below 80.00 (numbers 15 through 24) placed in Tier C.

Normally, when this type of system is used, the names of all candidates in Tier A are submitted to the appointing authority whenever a vacancy occurs. The appointing authority has the choice as to whom to hire. Tier B cannot be used until all those on Tier A (or perhaps a specified number) have been hired or justification has been presented and approved by the Civil Service Commission that one or more of the candidates should not be hired. When a vacancy occurs after Tier A has been exhausted, the names of all candidates in Tier B are submitted to the appointing authority. Tier C would be used after Tier B had been exhausted.

There are many variations of the Tier System. This system and the open system have been used by a number of jurisdictions for the position of firefighter and have also been used by many jurisdictions for promotional positions. In some jurisdictions that use this system for the position of firefighter, placement on the Eligible List is determined 100 percent by scores on the oral interviews or written examinations. The remaining portions of the examination process are graded on a pass-fail basis.

THE PROBATIONARY PERIOD

The probationary period is the last phase of the selection process. It is the working test part of the examination process. During the probationary period, those hired are observed and evaluated while performing the duties of the position.

Although it is designed as a working test period, the probationary time for entry-level firefighters normally includes some type of training. The length of the training period varies from one jurisdiction to another, with the longer period generally found in larger cities. Most large cities require that newly appointed firefighters satisfactorily complete the training in a well-organized department-operated fire academy prior to the recruit being assigned to a fire company. Unfortunately, most small communities have neither the facilities nor the staff to provide academy training for new recruits. The training takes place over a long period of time as an adjunct to the recruits' regular duties. In some parts of the country, however, the smaller departments are able to send recruits through fire academies conducted by community colleges or other educational institutions. Those departments that require that entry-level firefighter candidates be Firefighter II certified to take the examination are able to satisfactorily eliminate the training period. Hired individuals can be placed immediately on a fire apparatus, usually after receiving some brief training to learn the department's standard operating procedures.

Although training for newly appointed firefighters is provided in one form or another throughout the United States, training and professional development for newly appointed officers is sadly lacking. Only a few of the larger jurisdictions have extended training to this deficient area.

The minimum length of the probationary period is generally six months. Because of the length of the training period, some fire departments have established a minimum of one year for the probationary period. Two- and three-year periods, however, are often considered.

The probationary period is as much a part of the selection process as the written examination or the oral interview. While a candidate is receiving pay and performing the duties of the position, he or she does not have a permanent job. Candidates may be terminated during this period without recourse. In theory, all that is required is a report from a candidate's immediate supervisor that the member is not meeting the standards of the position. In practice, however, candidates are usually given as much assistance as possible in learning the duties of the position and in adjusting to the new environment. Few are terminated without the consensus of several superiors that the candidate's performance is not satisfactory.

TENURE

Once a firefighter recruit completes the probationary period, he or she becomes a permanent member of the department and is given tenure. Promotional members also receive tenure upon completion of the probationary period. Basically, tenure is a property right that cannot be taken away without due process of law. A firefighter, an apparatus operator, or an officer who has tenure is entitled to keep his or her position just as if it were his or her car, home, or anything else owned by the individual. Termination or reversion to a lower position requires that the employer follow established procedures for dismissal or reversion. These procedures are established by the civil service rules and regulations, codes, or ordinances. A firefighter or promotional member can, however, be terminated or reverted as a result of reduction in the work force. Termination or reversion procedures must follow due process of law and cannot be arbitrary or capricious.

SUMMARY

This chapter was written 100 percent in the informational mode. In it you were provided information you can use to formulate a solid foundation for understanding the examination process.

The Civil Service System was examined with an understanding of the spoils and merit systems provided. Each portion of the examination process was explored. An explanation was provided as to how an Eligible List is established and how candidates are hired from the list. Information was provided as to the differences that exist in the appointment procedure between the Rule of Three, the Rule of One, the Open Rule, and the Tier System. The difference in the status of a member on probation and one who has tenure was also examined.

Some of the remaining chapters will be written in the informational mode. However, most will be written in a combination informational/participation mode. Suggestions will be made in most of the chapters of things for you to do to prepare yourself for obtaining a satisfactory grade in an oral interview. It is important that you follow the suggestions in order to stay ahead of those with whom you will be competing for the number one spot on the Eligible List.

REVIEW QUESTIONS

1. What is the basis of the spoils system?
2. What is the concept of the merit system?
3. What is an Eligible List?
4. Which governmental body is responsible for ensuring that the principles of the merit system are enforced?
5. A candidate for firefighter believes he or she has been discriminated against during the examination process. To whom would he or she appeal?
6. What is a good method to use to remain aware of when different jurisdictions will conduct fire department entrance examinations?
7. In what places would an interested individual most likely find announcements of firefighters' examinations?
8. Why is it important to understand the entire examination process when preparing for an oral interview?
9. Where would a candidate for an entry-level firefighter's position find information that he or she could use to determine if he or she were qualified to compete in the examination process?
10. Where would a candidate who has one or more of the listed desirable qualifications most likely receive credit for them?
11. What are some of the various parts of a firefighter's entrance examination?

12. If it is used, what part of the examination process would most likely determine whether or not an individual will be hired?

13. What portion of the examination process generally determines a candidate's ability to perform the duties of the position?

14. If a practical phase is used in the examination process for a company officer position, where is it most likely to manifest itself?

15. A candidate for the position of firefighter receives a grade of 85.23 on the written examination, 91.65 on the oral interview, and 83.45 on the physical ability. The written is weighted 50 percent, the oral 35 percent, and the physical ability 15 percent. What would be the candidate's final grade?

16. What is generally the minimum passing score for a candidate to be placed on the Eligible List?

17. Approximately what is the maximum length of time that an Eligible List remains in effect?

18. What is meant by certification?

19. How many names are placed on the Eligible List if the hiring jurisdiction uses the Rule of Three?

20. How might the names be listed on the Eligible List if a jurisdiction is using the Open Rule?

21. A fire chief has an opening for firefighter. The city is operating under the Open Rule. There are forty-nine names on the Eligible List. How many names will be submitted to him or her from which a selection can be made?

22. What is the last phase of the selection process?

23. In general, what is the minimum length of a probationary period?

24. What does a probationary member receive when he or she satisfactorily completes the probationary period?

25. What is tenure?

BEFORE THE INTERVIEW

LEARNING OBJECTIVES

The objective of this chapter is to introduce a candidate to the number of things that he or she can do prior to the oral interview in preparation for completing in the oral interview process. As a result of reading this chapter, the reader should be able to:

- Explain how to learn about the duties and responsibilities of the position he or she is seeking.
- List the qualifications required for the position he or she is seeking.
- Explain some of the methods available for learning to present himself or herself effectively.
- List the duties of the position he or she is seeking.
- List ten things that he or she should know about the department prior to taking the oral interview.

INTRODUCTION

There are a number of tasks that should be completed before the interview. Some of these should be done immediately prior to the interview date, and others should be completed long before you receive notice to appear for the interview. The tasks that should be done after you receive notice to appear for the interview are included in Chapter 4. The others are listed in this chapter. You can help set yourself apart from the competition by carefully following the critical steps outlined in this and the following chapters. The tasks that should be completed for this chapter are divided into the following three categories:

1. Learn the duties and responsibilities of the position.
2. Learn the qualifications required for the position.
3. Learn to present yourself effectively.

LEARN THE DUTIES AND RESPONSIBILITIES OF THE POSITION

There are many common elements shared with the candidates who are seeking an entry-level position and those seeking a promotion. However, learning the duties and responsibilities for the position of entry-level firefighter normally takes more preparation than it does for a candidate preparing for a promotion. The reason is that those preparing for promotion have been able to observe someone performing the duties of the position for a number of years, while the duties of a firefighter are sometimes completely new to an entry-level candidate. This does not, however, mean that those preparing for promotion know all the answers regarding the position they are seeking and therefore do not need to explore the avenues open to them in order to learn as much as possible about the position.

> **Tips for Success** There are several methods that a person desiring to become a firefighter or to be promoted can use to learn as much as possible about the position. The importance of learning as much as possible about the position cannot be overemphasized.

One of the primary tools available to many candidates for learning about the position is the job announcement. However, the job announcement for some jurisdictions does not include any of this information or may provide only a limited amount. If it does provide a sufficient amount of information about the position, the information should be analyzed and the duties and responsibilities carefully examined. Following is a list of duties for the position of firefighter that was extracted from a job announcement for the position of entry-level firefighter for a medium-sized fire department. Although this description of the duties of a firefighter may not be exactly the same as those for a department to which a candidate may apply, it will give the candidate a basic idea of the duties of the position and the method that can be used to analyze the announcement of the examination for which he or she will be applying.

Examples of the Duties of a Firefighter

Responds to fires, emergency rescues, emergency medical care incidents, and public service calls; lays hose lines; operates apparatus pumping equipment as required; assists in holding nozzles to direct fire streams on fires; raises, lowers, and climbs ladders; conducts salvage and clean-up operations; operates resuscitators and other rescue equipment; administers emergency medical care; assists in the repair and maintenance of fire equipment and apparatus; answers general questions from the public; assists in maintaining a fire station; studies fire department rules and regulations, fire hazards, and firefighting techniques; participates in fire drills and exercises; paints, flushes, cleans, and performs minor repairs to fire hydrants; paints and checks fire alarm boxes; monitors and tests the fire alarm system in the fire station; participates in pre-emergency planning by inspecting and drawing floor plans of commercial, industrial, and related structures; acts as the relief engineer as needed; may be required to respond to fire calls during other than normal duty hours; and does other related work as required.

Now the announcement can be analyzed and the duties broken down into basic functions. From the announcement, the duties of a firefighter for the given city include:

- responding to alarms
- participating in emergency operations (see Figures 2-1, 2-2, and 2-3)
- administering emergency medical care (see Figure 2-4)
- repairing and maintaining fire equipment and apparatus (see Figure 2-5)
- participating in public relations
- maintaining the fire station (see Figure 2-6)
- studying
- participating in training exercises
- participating in pre-emergency planning and inspections
- maintaining fire hydrants
- maintaining fire alarm boxes
- monitoring and testing fire station alarm systems
- serving as the relief driver pump operator (see Figure 2-7)
- doing related duties as required

Figure 2-1 Engine company members use hose lines to attack the fire.

Figure 2-2 At large fires, engine company members occasionally operate heavy stream appliances.
Courtesy of Rick McClure, Los Angeles Fire Department.

Figure 2-3 Truck company members use axes and chainsaws to ventilate the roof.
Courtesy of Rick McClure, Los Angeles Fire Department.

Figure 2-4 One of the duties of a firefighter is to respond to emergency medical incidents.

Figure 2-5 Firefighters maintain the apparatus and equipment.

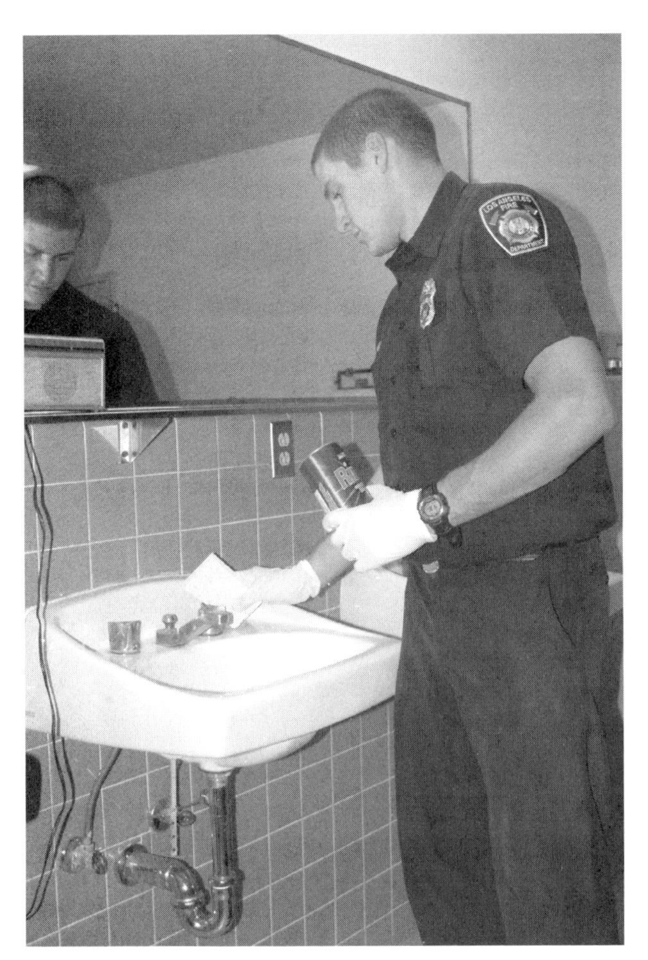

Figure 2-6 Maintenance of quarters is a responsibility of the firefighters.
Courtesy of Rick McClure, Los Angeles Fire Department.

Figure 2-7 Firefighters serve as relief apparatus operators.
Courtesy of Rick McClure, Los Angeles Fire Department.

It is important that an entry-level firefighter candidate learn as much as possible about the listed duties and be capable of organizing them into an effective presentation.

Examples of the Duties of a Company Officer

The same procedure can be used to analyze the duties of a company officer. The following was extracted many years ago from the "duties" of the job announcement for the position of captain in a large city. The city did not at the time use the oral interview as part of the examination process; however, the job announcement is used here as an example because it is very comprehensive. The original announcement has been modified for clarity and updating.

> Under the general supervision of a fire officer of a higher rank, responds to all emergency alarms in an assigned area and supervises all operations until relieved by a superior officer

> in order to bring the fire to a successful conclusion as soon as possible; responds to emergencies of a life-threatening nature and to other miscellaneous situations in order to save or reduce the loss of property or to assist people that are not necessary in a life-threatening situation; supervises, evaluates, and trains personnel in assigned daily operations of the company to ensure that all fire department methods, procedures, and daily routines are being carried out; evaluates personnel and equipment to determine the availability and readiness for emergency calls; instructs or arranges fire training for subordinates in order to keep subordinates abreast of new procedures, to maintain needed skill level of subordinates for efficient operations, and to develop subordinates to their fullest capability for promotion; schedules and assigns subordinates to various daily tasks that need to be performed in quarters; performs various supervisory functions required in supervising subordinates to develop a disciplined, effective, and cohesive unity while striving for a high level of morale; completes daily, weekly, monthly, and annual reports and forms required to keep accurate records of all actions performed by the company for the purpose of documenting such actions; supervises and performs inspectional, investigative, and regulatory duties pertaining to the prevention or extinguishment of fires; supervises and performs inspectional duties pertaining to pre-incident planning and fire prevention for all buildings in an assigned area; supervises routine in-service inspections, and investigates all complaints received; performs investigative duties to determine the cause of fires, false alarms, and other emergencies and public complaints when serving as the incident commander; performs various administrative functions as required; plans operations for which the company is responsible; attends and participates in meetings to solve various problems or to institute new procedures; performs public and community relations by means of attending community meetings and giving talks and demonstrations on fire prevention, first aid, and so on; performs related work as required.

An analysis of the duties for the position of a company officer from the preceding description will provide a workable list of duties that should be considered for an oral interview. Although the list encompasses only the duties as described in the illustrated examination announcement, it probably will apply to a company officer's position in most cities and communities throughout the country. Regardless, the analysis does illustrate how the duties can be determined for a community to which a candidate applies:

- responds to all emergencies in an assigned area (see Figure 2-8)
- supervises all operations at emergencies until relieved by a higher-ranking officer
- supervises personnel in the daily operation of a company (see Figure 2-9)
- instructs and trains personnel (see Figure 2-10)
- completes and maintains records (see Figure 2-11)
- determines the availability and readiness of personnel and equipment
- supervises pre-incident planning and fire prevention activities
- investigates and determines the causes of fires
- investigates public complaints
- performs administrative functions
- performs public and community relations
- institutes new procedures

Figure 2-8 Responding to fires and other emergencies is one of the primary duties of a company officer.
Courtesy of Rick McClure, Los Angeles Fire Department.

- presents and gives demonstrations to public groups
- does related work as required

A company officer candidate should be thoroughly familiar with each of the listed duties and be capable of organizing them into an effective presentation.

Following is an example of some of the questions a company officer candidate should be able to answer about his or her department prior to taking the oral interview:

- What records have to be maintained at the company level on a routine basis?
- What reports have to be forwarded on a monthly basis?
- What reports have to be forwarded on an other-than-monthly basis?

Figure 2-9 Directing a company at fires and other emergencies is the responsibility of a company officer.
Courtesy of Rick McClure, Los Angeles Fire Department.

Figure 2-10 One of the primary responsibilities of a company commander is to provide the training for company members.

Figure 2-11 A company officer's duties involve paperwork using modern technology equipment.

- How often do subordinates have to be evaluated, and what report is used for the evaluation?
- What disciplinary action can a company officer take without the approval of a superior officer?
- How do the duties of officers of the same rank on special assignments differ from those assigned to firefighting companies?
- How often must training sessions be conducted?
- What routine or periodical meetings must a company officer attend?
- Where in the department, other than to a fire company, can an officer of the position you are seeking be assigned?
- How often are pre-incident planning inspections required to be made?

A company officer candidate should place each of the listed questions on a 3×5-inch card for use in the practice sessions. A small portable file box should be purchased for use in storing the developed cards and those that will be added as progress is made throughout the book.

LEARNING BY LISTENING AND OBSERVING

Tips for Success The best method available for learning as much as possible about the job is to talk to individuals who are actively doing the job.

Entry-level firefighter candidates should visit fire stations and talk to the firefighters. They should find out what the firefighters do on a routine basis and information regarding other duties that are performed periodically. It is important that a candidate learn as much as possible about the daily schedule, the work schedule, and what takes place on special duty assignments.

A candidate should not limit his or her visits to a single station or a single type of district. Instead, stations that have various types of apparatus and stations in districts where the fire and emergency problems may be different should be visited. If the department operates fireboats, helicopters, or other special equipment, a candidate should be sure to visit these stations and learn as much as possible about the duties of the firefighters who operate this special equipment. A candidate should make sure to visit a station in a residential area, another in a commercial area, and others in special sections of the community such as brush areas, airports, and so on, if such areas exist. The more a candidate can find out about the job and the department, the more confident he or she will be if asked in the oral interview what a firefighter does.

Following is a sample of some of the questions that a candidate seeking a position with a particular city or community should be able to answer prior to taking the oral interview. Each of these questions should be placed on a 3×5-inch card for use in the practice sessions. A candidate should also purchase a small portable file box for use in storing the developed cards and other cards that will be developed as progress is made in the book.

- What is the work schedule for firefighters?
- How many hours a week do the firefighters work?
- Do firefighters participate in pre-incident planning inspections?
- Do firefighters make fire prevention inspections?
- Are firefighters required to participate in physical training every day they are on duty?
- Do firefighters test hydrants?
- Do firefighters test the alarm system in quarters?
- What type of equipment do firefighters maintain?
- Are all firefighters required to be EMT (emergency medical technician) qualified?
- Who does the cooking in the fire station, and how do members pay for meals?
- In what type of training activities do firefighters participate?
- How long is the probationary period for a new firefighter?
- Who pays for the uniforms a firefighter wears?

- What type of extra work are firefighters required to do when on probation?
- What time do firefighters report for duty?

A candidate seeking a promotion to company officer should not be remiss in this portion of his or her preparation. While candidates do have an opportunity to observe an officer performing the duties of the position, they do not really see everything he or she does. There are many reports required to be forwarded and many things that go on behind closed doors that remain unseen. A candidate should talk to a number of officers to find out exactly what their jobs entail. It is important for a candidate to visit those portions of the department with which they are not familiar and talk to officers responsible for these commands. Remember that a company officer can be assigned to many places in the department other than to a fire company. He or she should talk to some of these officers to learn if their duties and responsibilities are completely different from those of an officer in a fire station. What is discovered may be a complete surprise.

LEARN THE QUALIFICATIONS FOR THE POSITION

Up to this point, much of what has been written has examined what a candidate should do. From this point on, it will be considered that you are the candidate.

After the job has been analyzed, determine what knowledge and abilities are required to effectively do the job. Using the list of duties previously given for an entry-level firefighter, you should be able to establish, to some degree, the knowledge and skills necessary to perform the required tasks. Some communities hire certified firefighters; however, most hire individuals who have not participated in any type of training for firefighters. The following list includes some of the qualifications that might be required to do the job effectively. You

- must be physically fit (see Figure 2-12),
- should be capable of working with your hands,
- should be capable of presenting yourself effectively before a group,
- must be capable of learning,
- must be capable of working as a member of a team.

You might be able to make additions to this list in your evaluation of the position; however, this list should provide you with a base that will be used by the oral interview team when evaluating candidates for the position. If the final list you compile includes the abilities required to do the job, then the oral board will undoubtedly be asking questions of a candidate and observe his or her demeanor to see if he or she possesses these qualifications.

Promotional positions can be analyzed in the same manner. It is possible to use the list of duties previously examined for the position of company officer to develop at least a

Figure 2-12 Firefighting candidates must be physically fit.

partial list of the qualifications required for the position. An analysis of the duties and responsibilities as presented might produce the following. A company officer should have

- the ability to manage a company,
- the ability to effectively supervise people,
- firefighting knowledge and the ability to apply it,
- an adequate education that includes knowledge of the fire code and other related codes,
- the ability to teach,
- the ability to work effectively with people,
- the knowledge and ability required to prepare reports and maintain records,
- the knowledge of applicable laws,
- the knowledge of governmental agencies that affect his or her job.

Know the Department

It is important that you do not go into an oral interview without knowing as much as possible about the department to which you will be applying for a job.

> **Tips for Success** Knowing as much as possible about the department is important not only for candidates for an entry-level position, but also for those seeking promotion.

If you really want to become a firefighter, and you are applying for a job with a particular community, it is extremely important for you to learn as much as possible about that community and the fire department prior to opening the door to appear before the oral interview board (see Figure 2-13). If the community has an Internet Web site, much of the information you will need to know about the community may be available there. You will have to identify information that is not available on the Web site by other means.

One of the methods suggested is to visit the city hall to see if there are flyers available that will provide information about the community with regard to population, assessed valuation, department heads, and the governmental organization of the community. It is also important to visit fire department headquarters and learn as much as possible about the organization of the department. What is the name of the fire chief (see Figure 2-14)? What is the work schedule? How many personnel are on the department? How many fire stations does the department have? How would you define the department's value system?

While you are visiting fire stations to learn as much as possible about the job, you should also learn about the types of companies in the department, the special equipment available, the types of and number of fires or other emergencies the department responds to each year, and other information that might prove valuable if asked about it by an interviewer. It is also important to know the collar insignias that identify company and chief officers. Do not depend upon any information you know about the insignias used by other departments. There are no standard collar insignias used throughout the United States.

Figure 2-13 Learn as much as possible about the fire department.

Figure 2-14 What is the fire chief's name?

A Participation Assignment for Entry-Level Firefighter Candidates

Following is a sample of the questions you should be able to answer about the department where you plan to seek employment. Each of these questions should be placed on a 3×5-inch card to be added to those in the card file you bought for use in the practice sessions.

- Who is the fire chief?
- What collar insignia does the fire chief wear?
- How many stations are there in the department?
- How many engine companies are there in the department?
- How many truck companies are there in the department?
- What collar insignia do company officers wear?
- Other than engine and truck companies, what other types of companies are there in the department?

- Approximately how many alarms does the department respond to each year?
- Does the department respond to emergency medical aid alarms?
- What are the functions of an engine company?
- What are the functions of a truck company?
- What is the approximate population of the city?
- What is the assessed valuation of the city?
- How many people are in the fire department?
- Does the department have an automatic mutual aid agreement with any of the surrounding communities?
- How would you define the department's value system?
- What does the future hold for the fire department?

Those seeking a promotion should follow the same procedure. They should know as much as possible about the department budget, the plans for expansion, the department's training program, and other facets of the department. A visit should be made to the office of planning and research to find out where the department is going in the next few years. Visits should be made to other special-duty sections such as training, the fire prevention bureau, the alarm bureau, the mechanical repair shops, and other areas of the department. A candidate should talk to as many individuals as possible who are in a position to know what is happening and what is going to happen in the next few years in the department. It is important for you as a candidate to strive to know as much as any candidate with whom you will be competing for the job knows, and then a little bit more.

A Participation Assignment for Company Officer Candidates

Following is a sample of some of the questions you should be able to answer about your department if you are seeking a promotion to company officer. Make a 3×5-inch card for each of these questions and add the cards to those you previously made and placed in the card file you purchased for use in the practice sessions.

- What is the size of your department budget?
- What improvements are recommended in the budget?
- What are the department's short-term goals? What is the department's five-year plan for expansion (see Figure 2-15)?
- What improvements have been made during the past few years as a result of technological advances?
- What changes are expected in the following areas in the next five years?
  - training
  - fire prevention
  - alarm bureau
  - mechanical shops

Figure 2-15 Are you knowledgeable about both of these?

- apparatus
- firefighting tactics
- emergency medical care
- organization

LEARN TO PRESENT YOURSELF EFFECTIVELY

Although the ability to present oneself to a group may seem to come naturally to some, for most people it requires a great amount of work. In addition to the need to overcome a seemingly inborn tendency to become nervous, there is also the need to learn to organize one's thoughts and present them in a manner that will attract and hold the interest of others. This task not only requires that an individual develop the ability to think on his or her feet, but also that a person develops, to a limited degree, the capabilities of an overdramatic actor. It may sound impossible; however, all is not lost. By taking some known paths, help is on the way.

Tips for Success There are several methods you can use to learn to present yourself effectively before a group. These methods will also be useful when talking on a one-to-one basis. The two most often used methods are courses in public speaking or oral interviewing at colleges and joining a **Toastmasters** or **Toastmistresses** group.

Of the two, joining a Toastmasters or Toastmistresses club is by far the most effective. The reason is that a Toastmasters or Toastmistresses Club has a well-designed program that is used for the professional development of members. On the other hand, a public speaking class in a college provides experience in public speaking, but the course is not standardized and the objectives and results depend a great deal upon the qualifications and experience of the instructor.

Toastmasters and Toastmistresses clubs are organized and designed to assist individuals in their professional development by teaching them how to better prepare themselves for public speaking. Members of Toastmasters and Toastmistresses clubs come from all walks of life. Attendance at a meeting will find teachers, salespersons, police officers, firefighters, bank presidents, union officials, industry workers, and blue-collar workers from various trades. They are gathered together because of a common need—the need to learn to communicate effectively. Painful as the learning may seem, the people gathered together recognize that one of the basic needs for survival and advancement in most occupations cannot be learned from a book. Just as in learning to swim, it is necessary to get into the water and get wet. Toastmasters and Toastmistresses own the pool. When asked, they will push you in.

These clubs are found in most major cities. In fact, in some of the larger communities, clubs are organized and thrive within large industrial organizations. The clubs might be listed in the telephone book, or, if not, the local chamber of commerce generally maintains a list of clubs together with the names of club presidents. If there is no club within your community, information is available on the Toastmasters and Toastmistresses Web sites on the Internet as to the location of the nearest clubs to your community. It generally takes only a call to the president to be invited to a meeting.

Some clubs hold breakfast meetings, some luncheon meetings, and some supper meetings. Most clubs meet weekly; however, a few gather together ever other week. Dues normally include the cost of the meals. A member can expect that for one reason or another, the club will have him or her on his or her feet at least twice during every meeting. One of the reasons will be because the member is given a couple of assignments for the meeting.

Assignments are generally divided into three groups: (1) special assignments, (2) prepared speeches, and (3) impromptu speeches. If a member is given a special assignment, he or she may be the Toastmaster or Toastmistress for the evening and be responsible for the overall operation of the meeting. In addition, a member may be assigned as an evaluator for either the prepared talks or the impromptu talks, or perhaps be assigned as the grammarian. Each of the evaluation assignments require that the member assigned listen very carefully for organization, methods of gaining attention, and general effectiveness of the presentation. After all the presentations have been made, the evaluator offers his or her observations to the group, pointing out the effective portions of the presentations and those areas that could be improved. The assignment as an evaluator is excellent training for learning to evaluate oral interview preparations and presentations.

The prepared speeches are normally five to seven minutes in length. A course available to members includes various types of prepared speeches. Included are serious talks, humorous presentations, and speeches that are read. An individual is evaluated on his or her presentation and, over a period of time, will learn what works and what does not. This part of the program is very similar to the one an oral interview candidate will have to develop for himself or herself in preparation for answering certain questions that he or she knows will be asked by the oral board. Of course, in the final analysis, the important part will not be in the preparation of an answer but in the ability to be oneself when making the presentation.

The impromptu part of the program probably approaches most closely the major portion of an oral interview. The individual who is responsible for conducting this portion of the program might say, "And now it gives me great pleasure to introduce Mr. Tim Smith, who will give us a presentation on the natives of Outer Mongolia." From the time Mr. Smith leaves his seat to the moment he arrives at the podium, he has to prepare a one- to two-minute presentation on these elusive natives. Of course, if Mr. Smith is extremely familiar with the lifestyle and habits of these people, he might not have too much trouble. However, it is extremely unlikely that this is the case. This then presents a real challenge. Most new members are pleasantly surprised at how well they are able to meet the challenge after a few months in the club.

The practice received in the club will help an oral interview candidate prepare for those unexpected questions that are thrown at him or her in the interview. A candidate will gain experience in thinking on his or her feet and learn to organize and present his or her thoughts in a logical manner in a short period of time.

In addition to their regular program, Toastmaster and Toastmistresses clubs try to rotate every member through the club officer positions in order to assist in developing leadership capabilities. A member starts out at the bottom and works up to the position of president. This progress gives a member valuable experience in every facet of a club's organization. This is extremely valuable when a firefighter or promoted member becomes involved in civic functions. A member is also exposed to and learns a considerable amount about parliamentary procedure and *Roberts' Rules of Order*.

The best advice that can be given at this point is, "Don't wait." Join a Toastmasters or Toastmistresses group today. It is an important part of the training and professional development that is needed in the preparation to take an oral interview. It is also part of the training you will find extremely interesting and rewarding. Once you become a member, a door opens that will guide you to learn to speak better, listen more carefully, and think better on your feet.

Joining a Toastmasters or Toastmistress club or taking a class in public speaking at a community college can be considered the theoretical portion of learning to present yourself well before a group. It is true that in either situation you will receive the benefit of talking to a group. However, the people you will be talking to are there more to improve their own proficiency than to listen to you talk.

Tips for Success For maximum benefit, it is best that you also include some practical experience in your learning atmosphere.

The practical experience is gained in talking to a group that is there to listen to what you have to say. It could be a civic organization, a social group, or perhaps a church group. Regardless of the composition of the group, the experience you gain will be beneficial to your overall goal.

Tips for Success Don't shy away from your chance to talk before a group. Seek the opportunity, and take it every chance you get.

SUMMARY

This chapter is an informational/participation chapter. This is the first of nine chapters that have been organized and designed to take you down a path from where you are to where you would like to go. Along the way you will be faced with a number of challenges. However, none of them are insurmountable. If you do your homework and learn to present yourself effectively, your goal is obtainable. However, if you allow several of your competitors to better prepare themselves, you will most likely fall short of obtaining your goal.

In this chapter you have been provided information on how to determine the duties and responsibilities of the position you are seeking, and how to learn as much as possible about the department to which you are applying for a position. You have also been provided information on how best to learn to present yourself effectively. You have not completed this chapter until you have enrolled in a course in public speaking or have joined a Toastmasters or Toastmistresses club; have gathered the information on the duties and qualifications of the position you are seeking; have gathered the knowledge about the community and fire department where you will be applying for a position; and have organized the information into effective presentations for answering questions on these subjects asked by the oral interviewers. It is only then that you can consider this chapter satisfactorily completed.

One of the presentations you are asked to prepare as a result of reading this chapter is one of the most important considered in this book. The presentation is an answer to the question, "What are the duties of a firefighter or company officer?" This question will most likely be one of those referred to as an opening question. Your response and presentation to the opening questions are the most important of the entire interview.

REVIEW QUESTIONS

1. List ten duties of a company officer from the job announcement given in this chapter.
2. List ten duties of a firefighter from the job announcement given in this chapter.
3. List the three categories of tasks that should be done before your oral interview.

4. What are the primary tools you have for learning about the position you are seeking?
5. Other than the examination bulletin, what is the best method for learning about the duties of the position you are seeking?
6. What is probably the best method of learning to present yourself effectively?
7. What kind of presentations are members of a Toastmasters or Toastmistresses club required to give?
8. What is an impromptu talk?
9. What are some of the avenues available for a candidate to improve his or her ability to present himself or herself?
10. You think you will be notified to take an oral interview examination in six or seven months. How soon should you join a Toastmasters or Toastmistresses club?

THE FIRST IMPRESSION: APPLICATIONS AND RESUMES

LEARNING OBJECTIVES

The objective of this chapter is to explain to the reader the primary factors that are used by oral board members to form a first impression of a candidate. As a result of reading this chapter and gathering the information recommended, the reader should be able to:

- Complete an application for employment.
- Develop, organize, and prepare a resume.

INTRODUCTION

How many times have you met someone you liked immediately, but later when you became much better acquainted, you disliked him or her intensely? Or the reverse may have happened. You may have met someone whom you disliked at the initial contact, but later the two of you became the best of friends. It is a known fact that many married couples fall into this latter category. First impressions are important, but first impressions are not necessarily lasting. First impressions cannot, however, usually be changed in twenty minutes. Consequently, the first impression you make on the members of the interviewing board is extremely important as it will undoubtedly have some effect upon the grade you receive.

THE APPLICATION

Many people are of the opinion that the first contact you make with the oral board members occurs when you are taken into the room and introduced to the chairperson of the board, and that the first impression you make is based upon your appearance and your manner at the time when introductions are made. While this contact and impression is extremely important, it is usually the second impression, not the first. In most cases, the board has reviewed your application and resume (if you have submitted one) for at least five minutes before you were taken into the room. These papers are their first impression of you, and it is extremely important that they be a favorable one. Yet applications are submitted with misspelled words on them, requested information omitted, and the application not signed. An example came from a candidate who had majored in psychology in college, and misspelled the word "psychology" on his application. It is not difficult to judge the first impression he made. When questioned about the misspelling, he stated that a friend had typed the application for him and the individual probably did not know how to spell "psychology." It did not seem wise to any of the oral board members for a candidate to trust his professional career to a friend who did not know how to spell. It was even worse that he had not checked the application and caught the error.

Applications are important. They are a reflection of your professionalism, your organizational ability, your thoroughness, your interest in the job, and your general attitude.

Tips for Success Make sure your application reflects a good picture of you and that it will leave the oral board with the best possible first impression.

Here are eleven points to check that can be advantageous to you.

1. Have your application professionally reviewed for spelling, grammar, sentence structure, and so forth. Although the word processor on your computer will check for spelling and grammar, it too can make mistakes. Do not put 100 percent trust in it.
2. Have your application professionally typed. Do not trust this to a friend who might have a computer or typewriter that does not do a professional job.
3. Some communities require that the application be printed in black ink. If you do not print well, have someone do it who does. However, make sure you thoroughly check the application before turning it in.
4. Make sure the application is complete. If information is requested that does not apply to you, do not leave the space blank. Write in the words "not applicable," or place a dash in the space provided. If you do not, the oral board will not know whether you neglected to complete the application or if the information does not apply. Be particularly careful to ensure that your work history is complete.
5. Where possible, center words. As an example, where your name is to be written, center your first name over the word "first," your middle name over the word "middle," and your last name over the word "last." Use this process throughout the application.
6. Make sure all marks are in the center of the box provided if the application asks you a question such as if you are a veteran and you indicate your answer by placing an X in a box. Be sure the X is exactly in the center of the box, not off to one side or in a corner.

7. Use full names of organizations the first time they are used, not abbreviations. For example, use National Fire Protection Association, not NFPA. This point is particularly important if a resume is not submitted with the application. Remember, one or more members of the oral board may not know what the letters NFPA stand for.
8. Use the full address and zip code for every residence in which you have lived for the past ten years. Also use the full address and zip code for every former employer and reference listed on the application.
9. Make sure you follow all the instructions regarding the application. Some jurisdictions will require that you submit a photo or some other unique item as part of the application process. Do not neglect to attach these items to your application.
10. Be sure to date and sign the application. A number of jurisdictions will not even accept an application that is not signed.
11. Some jurisdictions will accept an application on the Internet. All the above information applies. Have someone who is professionally skilled type out the information for you. However, you should sit alongside him or her to make sure all the information submitted is absolutely correct.

Here is one last word about applications. Be sure to keep a copy and also store it on your computer and on a computer disk so it can be updated easily. Much of the information that will be requested in future applications will be the same as that for the application you are submitting. Normally, all you need to do is add new information to the present application. Keep a file somewhere with your application in it. You will be surprised at how many times you will need it for reference.

As a matter of fact, the first time you will have to refer to it is shortly after you complete it. Remember that the oral board reviews your application before they see you. Your application will serve as a source of questions they might want to ask you. Be sure to review the application shortly before the interview. Remember, with your application you have opened a door. Be prepared to answer any questions that might be asked because of it.

RESUMES

Every candidate who submits an application for a company officer position should submit a resume. Some candidates for an entry-level firefighter position should submit one and others should not. Whenever this statement is made in a class on oral interviews, someone almost always questions it. "Why," the individual asks, "should you submit a resume when the oral board has your application? Don't you put the same information on the resume that you have on the application?" The answer, of course, is "yes and no." To explain this answer, it is first necessary to understand the difference between an application and a resume from an applicant's point of view.

An application provides the oral board with information they may want to know in a format chosen by the examining jurisdiction.

A resume provides the oral board with information a candidate wants the oral board to know in a format the candidate has chosen and organized so that he or she can make the best impression possible. It also provides the candidate with a means of guiding the oral board into asking questions he or she wants them to ask. What could be better?

To achieve the objectives you are striving for, consider the following.

Critical Points to a Good Resume

- Lay out the resume in a form that will get the interviewers' attention quickly.
- Limit the resume to one page (see Figure 3-1). Remember, the resume is a summary of yourself, not a life history. The interviewers have only about five minutes to review both your application and your resume. Do not defeat your purpose by giving them so much to read that they cannot do it in the limited time available.
- Provide adequate margins and leave plenty of white space. Make the resume as attractive as possible.
- Use a good grade but not fancy paper. The difference in cost between standard paper and a good grade is minimal. You do not want any other applicant's resume looking better than your own.
- Do not tell the board everything on your resume. Give them just enough information so that they will ask you questions concerning those things you want to talk about.
- Seek professional help, if necessary, to ensure that your resume presents the best picture possible.
- Make sure all words are spelled correctly and that everything in the resume is grammatically correct. The computer will assist you in this effort, but don't depend entirely on it.
- Have your resume professionally typed.
- Use the past tense throughout the resume, even if it involves something in which you are currently involved.
- Use action verbs. Some of the action verbs that are closely associated with an officer's duties and responsibilities are: planned, organized, directed, coordinated, and controlled. Some other action verbs that might be beneficial in a resume are shown in Table 3-1.
- Never use personal pronouns on a resume.
- Never include personal data that is irrelevant to the job, such as marital status or religion.

Tips for Success A resume should be constructed in such a manner that it provides the board with a mental image of you in regard to your back ground, experience, and qualifications for the position. It should stress the positive.

Resume of background and experience

Jim Doe

Supervisory and Management Experience

1. Acting Captain, three years—In the absence of the captain, planned, organized, directed, coordinated, and controlled the activities of a fire company.
2. Supervision and Management courses—Completed a course in supervision and one in management. Learned the principles of supervision, management, discipline, and training.
3. Past President of Toastmasters—Learned the principles of presenting material to a group and the principles of planning and organizing group functions.

Firefighting Experience

1. Engine company experience, six years—Learned the principles of size-up, estimating water requirements, hose placement, and tactics for various types of structure fires, transportation fires, wildland fires, and fires in different types of occupancies.
2. Truck company experience, two years—Learned the auxiliary functions of laddering, overhaul, ventilation, forcible entry, physical rescue, salvage, and control of utilities.
3. Fireboat experience, one year—Learned waterfront attack tactics and shipboard firefighting tactics.
4. Completed courses in firefighting tactics, hazardous materials, and wildland fire control at a community college. These courses strengthen the experience of practical field operations.

Training and Education

1. Associate of Science degree with major in fire science—Anytown Community College
2. Working toward a B.S. degree in Fire Administration at State College
3. Completed courses in instructor training at State Fire College. These courses will be beneficial in the development of a company training program.
4. Taught a course in firefighting tactics at Anytown Community College. This experience will be beneficial in planning and organizing company training.
5. When acting captiain, prepared lesson plans and presented training sessions at the company level.

Related Experience

1. Scoutmaster, Boy Scouts of America—Planned, organized, and directed boys in scouting and related activities.
2. Member of Anytown Firefighter's Association—Served one year on negotiating team. Learned the principles of give-and-take when differences of opinion exist.
3. Married, three sons—Learned budgeting, discipline procedures, and necessity of give-and-take in group relations.

Figure 3-1 A sample resume for a company officer.

TABLE 3-1 Use action verbs for a good resume.

administered	attended	controlled
analyzed	completed	converted
arranged	conceived	coordinated
assigned	conducted	correlated
created	inspired	programmed
delegated	installed	proposed
demonstrated	instituted	recorded
designed	integrated	researched
determined	interviewed	rewrote
developed	investigated	scheduled
directed	maintained	served
documented	managed	simplified
established	monitored	solved
evaluated	motivated	strove or strived
examined	negotiated	supervised
executed	operated	taught
formulated	organized	trained
implemented	participated	updated
improved	performed	worked
increased	planned	wrote
initiated	prepared	
inspected	processed	

While all candidates, including yourself, have a few weaknesses with respect to their qualifications, none of yours should manifest themselves either directly or indirectly on your resume.

Keep in mind that a resume is an individually designed tool that summarizes your background, education, experience, and qualifications for the position. It should be developed in a manner that paints the best possible picture of you. Of all the people who read it, it should please you most. If you ask several people to review it, at least one will probably be critical of some portion of it. Some may say it is too professional. Others may complain that it is poorly organized or does not provide sufficient information. These reviewers generally reflect an opinion based upon their own backgrounds and lifestyles. Listen to them, but do not make changes on the basis of their opinions only. It is your resume. If it reflects what you want it to say, and if you are satisfied, proceed with it. Of course, if all your reviewers make the same negative comments, it is probably worthwhile for you to take another look at what you have put together.

When your resume is complete, review it for some of the common mistakes made on resumes. If you use the suggestions made earlier as a checklist, such mistakes should be minimal.

Common Resume Errors

- poor organization
- spelling errors and poor grammar
- poor grade of paper
- poor typing
- valuable experience omitted
- information presented in a confusing manner
- unimportant information that is not relevant to the position sought
- submission of a resume designed for a previous examination and not updated

There are two basic types of resumes: (1) a **chronological resume** and (2) a **functional resume.** A chronological resume highlights skills, training, and experience chronologically over a period of time. A functional resume highlights a variety of skills and qualifications that are relevant to the position sought. It is generally better to use a functional resume for both the entry-level and promotional level of the fire service. With a functional resume, you take all the facts you want to include in your resume and list each one under one of four or five functions. The functions selected will depend upon your particular strengths and the picture you want to convey. For example, five functions you may wish to consider for the position of company officer are: (1) supervision experience, (2) firefighting experience, (3) training experience, (4) education, and (5) special assignments and related activities. When you select these functional divisions, be sure to define in your own mind exactly what you mean by the words. For instance, you will find many definitions of the word "supervision." Some people have thought of "supervision" as incidents in which they directed, controlled, and consequently got their work done directly through other people. These same people may have separated "supervision" from "management" by thinking of management functions as those in which they had planned, organized, and controlled the activities of a group. What do you think? If you are going to use the terms "supervision" and "management," how would you define each of the two?

Some of the various factors you might want to consider for inclusion in your resume appear in the list that follows. A few of these could be used as a functional division, while others should be included within a division.

Entry-Level Position

In addition to the possible separate functions of work experience and education, an individual or candidate applying for the position of firefighter should consider including categories within a functional division that demonstrate the ability to

- apply knowledge learned,
- get along well with people,
- improvise,
- learn,
- present oneself efficiently,
- read and understand written instructions,
- remain calm under stress,
- think and act quickly,
- understand and follow orders,
- work as a member of a team
- work with one's hands.

Each of these candidates should also consider categories that demonstrate he or she

- has a sense of personal responsibility,
- is cautious,
- is dependable,
- has endurance,
- has good work habits,
- is a good worker,
- is honest,
- has initiative,
- is organized,
- has physical strength and coordination,
- has self-confidence
- is well-disciplined.

Company Officer Candidates

In general, the categories for a company officer candidate are entirely different from those of an entry-level firefighter; however, entry-level candidates should review the following list as some of the categories may apply to their situations. The possibility is particularly true for older candidates and those who have had military service.

- civic functions
- dedication and loyalty

- education (all levels: high school, college, and advanced degrees)
- elected positions held
- firefighting experience
- foreign language ability
- computer literacy
- honors
- inventions
- leadership experience
- licenses held
- professional certifications
- management experience
- personal interests
- presentations made
- professional affiliations
- publications
- research activities
- service organizations
- special duty assignments
- special skills
- staff experience
- supervision experience
- training experience
- volunteer activities

Most people who work in the private sector who submit resumes do not submit them together with an application for employment. These individuals prepare resumes and mail them to prospective employers in the hope of getting an interview. The objective of their resumes is different from those for a fire service position. While the objective of submitting a resume for an individual in the private sector is to obtain an interview, a fire service candidate already has one. Consequently, private sector resumes contain personal information such as address, telephone number, and so on. It is not necessary to add this type of information on a fire service resume, as the application generally provides this.

Figure 3-1 shows a sample resume for a company officer.

Resume Guidelines for the Position of Entry-Level Firefighter

Following is a suggested list of information to gather prior to writing your resume. Gathering this information will not only assist you in preparing a resume, but will also provide you with information you may need to answer questions asked by the interview board.

It is preferred that you list the answers on a piece of paper and then identify the function you will assign it to if you decide to use the information.

Remember, the resume should be limited to one page. Therefore, you will probably be gathering more information than you will be able to include on your resume. It is suggested that you prepare a question that you think the oral board may ask about this information and make a 3×5-inch card for use with the practice sessions. Add the cards to those in your card file.

- List your education. Start with the most important (that is, M.S., B.S., and so on). Include the name of the school, dates attended, major, and degree earned.
- List your work experience. Start with the present or most recent. Include the name of the company, dates, position(s) held, and an example of your duties. Remember to use action verbs.
- List three examples of incidents where you were able to apply some knowledge you had learned.
- List three examples of incidents where you displayed the ability to get along with people.
- Give three examples that indicate your ability to improvise.
- Give three examples that indicate you are capable of learning.
- List three incidents to demonstrate that you are capable of presenting yourself efficiently before a group.
- Provide three incidents that indicate you have the ability to read and understand written instructions.
- List three occasions when you demonstrated the ability to remain calm under stress.
- List three incidents where you demonstrated you have the ability to think and act quickly.
- List three incidents that indicate you have the ability to understand and follow orders.
- List three incidents that indicate you have the ability to work as a member of a team. Only one of the three should refer to sports.
- List three examples of your ability to work with your hands.
- List three incidents that indicate you have a sense of personal responsibility.
- List three incidents that indicate you are a prudent individual.
- List three situations that show you are a dependable person.
- List three activities in which you have participated that indicate you have endurance.
- List three situations that show you have good work habits.
- List three activities that indicate you are a hard-working individual.
- List three incidents that indicate you are an honest person.
- List three incidents in which you displayed initiative.
- List three incidents that indicate you are an organized person or have the ability to organize.

- List activities or incidents that show you have the physical strength and coordination to be a firefighter. Try to use only one sport activity or incident in your response.
- List three incidents that would indicate you have the self-confidence to be a firefighter.
- List three incidents that indicate you are a self-disciplined individual.
- List five functional divisions you can use on a resume to best indicate the strengths in your background and experience.
- List at least three factors under each of the above divisions to support its credibility.
- Using all of the previous information, prepare a resume for yourself.

Resume Guidelines for the Position of Company Officer

Following is a suggested list of information to gather prior to writing your resume. Gathering this information will not only assist you in preparing a resume, but will also provide information you may need to answer questions asked by the interview board. It is preferred that you list the answers on a piece of paper and then identify the function you will assign it to if you decide to use the information.

Remember, the resume should be limited to one page (see Figure 3-1). Therefore, you will probably be gathering more information than you will be able to include on your resume. Prepare a question you think the oral board may ask about this information and make a 3×5-inch card for use with the practice sessions. Add the cards to those you have in your card file.

- List your education. Start with the most important (that is, M.S., B.S., and so on). Include the name of the school, dates attended, major, and degree earned.
- List the three most important civic functions in which you have participated.
- List three activities or projects in which you participated that indicate you are creative.
- What are the three most important elective positions you have held?
- List all the firefighting experience you have had by type of company (engine, truck, other types). With each company, provide some specific examples of things you learned that have qualified you for the position of company officer.
- List five fires or emergency incidents in which you learned something that specifically helped qualify you for the position of company officer.
- What foreign languages are you capable of speaking, reading, or writing?
- List three of your hobbies. Indicate how any of these might help qualify you for the position of company officer.
- What honors have you been awarded?
- Have you invented anything? If so, do you feel that this contributes to your qualifications for the company officer position?

- List the positions of leadership you have held. Give specific examples of what you learned in each position that has helped qualify you to be a company officer.
- What licenses (other than a driver's license) do you hold? How do you feel that each of them helps qualify you to be a company officer?
- List each of the management positions you have held. Give specific examples of what you learned in each position that helps qualify you for a promotion.
- List three personal interests you have other than hobbies. Explain how any of these might be beneficial to you as a company officer.
- What three presentations have you made to a group of five or more people that you feel demonstrates your ability to present yourself well before a group?
- List your three most important professional affiliations and explain why your association with these organizations helps qualify you for the position of company officer.
- What articles, books, and so forth have you had published?
- List the research projects you have participated in and explain what you learned in each that would be beneficial to you as a company officer.
- List any special duty assignments you have had in your current or previous departments. List what specific things you learned in each of these assignments that will benefit you as a company officer.
- What special skills do you possess that will be beneficial to you as a company officer?
- List any staff experience you have had, other than your special duty assignments. What did you learn from this experience that will benefit you as a company officer?
- List your supervisory experience. For each position, give the number of people you supervised and list at least three specific things you learned that will be beneficial to you as a company officer.
- List the experiences you have had in training people. Include under this item courses you have taken to learn the principles of instructing others. Give specific examples of what you learned in each of these classes that will benefit you as an instructor.
- List your volunteer activities and explain what you learned from these activities that will be beneficial to you in managing a fire company.
- What five functional divisions will you use on a resume that will best illustrate your strengths for the position of company officer?
- List five activities under each division that will give the function credibility.
- Design and construct a resume that will best indicate your qualifications for the position. Keep in mind that a resume opens doors (see Figure 3-2).

Tips for Success Make sure your resume is organized in a manner that will guide the oral board to ask you the questions you wish to discuss.

Figure 3-2 A resume opens a door.

SUBMITTING YOUR RESUME

It is important for you to determine whether or not the jurisdiction to which you will be submitting an application will permit resumes to be provided to the oral board. If they do, you do not want to attach your resume to your application. An application may be handled by a number of people prior to reaching the members of the oral board. In the process, it could become dirty or wrinkled. You will spend a lot of time and effort in producing a resume that you hope can be used to give the oral board a good first impression of you. Don't goof it up.

If resumes are accepted, find out how many people will be on the oral board. Then have that many copies of the resume made plus one for yourself. Make sure that the copies are on the same quality of paper. Leave yours at home for future reference. Give those for the oral board to the receptionist who will be taking you into the oral board room and ask him or her to give one to each member of the oral board when the application is submitted to them. This will give every member of the board time to review your resume prior to your entering the room.

If resumes are not accepted, it is recommended that you still gather the information suggested to be gathered for the resume. This information will be valuable in answering oral board questions.

SUMMARY

This chapter has provided you with some of the important information you should consider in order to make a good first impression. Good impressions are critical to obtaining a good grade on your oral interview.

This chapter provided information for completing an application. However, the primary emphasis was placed on what information to gather and how to prepare it for completing a resume. The critical points for completing a good resume were discussed and the common errors made on resumes were listed. A sample resume was provided for consideration.

Some of the participation phases for this chapter were listed for completion as you read the chapter. However, this chapter cannot be considered totally completed until you have organized and prepared an outstanding application and resume ready to be submitted, and you have prepared the 3×5-inch cards suggested in the chapter.

REVIEW QUESTIONS

1. From what contact does the oral board form its first impression of you?
2. What factors does your application reflect?
3. What are the seven points you should check to ensure your application presents the best picture of you?
4. Why is it important to keep both a copy of your application and one on file in your computer?
5. When should you review your application, and why is it necessary to review it at this time?
6. Who should submit a resume with his or her application?
7. What is the difference between an application and a resume?
8. What should be your two primary objectives in submitting a resume?
9. List ten points for completing and designing a resume.
10. What are the five best action verbs to use on a resume for the position of company officer?
11. What should you do about weaknesses in your qualifications in regard to your resume?
12. What are some of the basic errors made on resumes?
13. What are the two basic types of resumes?
14. What type of resume is generally best for fire service candidates?
15. What type of personal information should you not include on a resume?

THE INTERVIEW PROCESS

LEARNING OBJECTIVES

The objective of this chapter is to introduce a candidate to the overall interview process. Upon completing this chapter, the reader should be able to:

- Explain the objective of the interview.
- Describe what to do when you receive notice to report for the interview.
- State what should be done while waiting for the interview.
- Describe how long the interview will most likely last.
- List the types of interview boards to be expected.
- Explain the types of instructions that are given to the board.
- Describe what the board knows about a candidate.
- List the types of questions a candidate can expect to be asked.
- State how the interview will most likely be terminated.
- Describe what takes place when the interview is over.
- Explain what a candidate can do if he or she feels that the board has discriminated against him or her.
- State what to expect in a pre-appointment interview.

INTRODUCTION

Being interviewed for a job is almost as inevitable as death and taxes. Anyone who enters the work arena will eventually have to come face-to-face with its reality. Some interviews are conducted on an informal, one-on-one basis; others are more formal. The objectives and principles involved, however, are basically the same. Today, at least 99 percent of American companies use the interview as part of their selection procedure, while nearly every governmental agency utilizes it in their promotional processes and also for their entry-level positions. Despite this fact, most candidates fail to attach to interviews the importance they deserve.

In the majority of cases, the fire department oral interview is the most important part of the examination process. Seldom is it given a weight of less than 30 percent, and in some jurisdictions a candidate's position on the Eligible List is determined 100 percent by the results of the interview. Consequently, the short time spent in the interview room may be the most important event in a person's work experience with an organization as the results may determine his or her entire future. Yet it is surprising how many candidates reporting for an interview have made little or no preparation.

Some candidates will spend months preparing for the written or practical portion of the examination process, but will devote only a few hours, or even a few minutes, thinking about the interview. They assume that because they meet the minimum qualifications as outlined on the examination bulletin that they are qualified for the position, or they have a preconceived idea that little can be done to prepare for the interview. Both of these assumptions are far from reality.

Having the minimum qualifications stated in the bulletin only allows a person to compete in the examination process. It is a fact that much can be done to prepare for the interview. Of course, you must be the type that believes that some preparation must be made or you would not be reading this book.

OBJECTIVE OF THE INTERVIEW

Most examination processes are designed to test a candidate's technical knowledge, his or her skills, if necessary, and certain intangibles required for success in the position. Technical knowledge is usually tested by a written examination, skills by a performance test, and the intangibles by an oral interview. While a person's training and experience can be outlined on an application or resume, the oral interview is the only place where the value of the training and experience and a candidate's personal qualifications can be measured. Yet in many respects, these factors often determine success or failure on a job. Contrary to common belief, more people are discharged from fire department positions because of a poor attitude or poor personal attributes than because of a lack of skill in job performance. It is therefore worthwhile to consider some of the common personal traits that may be measured by the oral interview. Although the following list is by no means complete, it should provide a candidate with some understanding of the intangibles involved:

- ability to organize and communicate thoughts
- ability to present oneself
- ability to work under stress
- ability to work with others
- ability to follow instructions
- adaptability
- attitude
- calmness

- emotional control
- enthusiasm
- honesty
- initiative
- innovativeness
- judgment and common sense
- leadership
- loyalty
- motivation
- personal appearance
- poise
- self-confidence
- stability
- tact
- tolerance of boredom
- value of education and training
- voice

RECEIVING NOTICE

The first positive indication that you will be invited to participate in the oral interview occurs when you receive notice as to the date, time, and place of the interview (Figure 4-1). When you receive the notice, write the information on a calendar and post the notice in a place that will serve as a constant reminder. Unfortunately, there have been candidates who have arrived at the interview location only to find that they reported a day early, or, worse yet, a day late.

NOTICE TO APPEAR FOR AN ORAL INTERVIEW FOR THE POSITION OF

____

DATE AND TIME: ______________________________

LOCATION: __________________________________

NO SUBSTITUTIONS OF DATE OR TIME WILL BE ALLOWED. If you are unable to keep this appointment, please notify us as soon as possible. Call (000) 000-0000

PLEASE BRING THIS NOTICE AND A PICTURE ID WITH YOU WHEN REPORTING FOR THE INTERVIEW

Figure 4-1 A sample notification card.

Prior to the date of the interview, travel from your home to the interview location, noting possible delay factors such as freeway traffic, railroad crossings, and so on. Write the time required to complete the journey on your calendar. While in the vicinity of the interview location, take the time to observe places to park and anticipate possible difficulties or delays. Remember that the availability of parking spaces will vary according to the day of the week and the time of the day. Traffic will be heavier when people are going to or returning from work, and a parking lot that may be almost empty on a Saturday may have a limited number of spaces when people are working in the area.

Check the parking lot to see if a given amount of change will be required for a meter. If so, you should be sure to have it on the day of the interview. In fact, it is better to put the required change in the glove compartment to ensure that it will be available when needed. Be aware that you will be nervous on the day of the interview, and any unanticipated delays will only add to your frustration.

In big cities, it may be necessary to travel to the interview location by public transportation. Make sure you know the time of departure and the arrival time. Write these times down on your calendar. It is best that you make the trip prior to the date of the interview so that you are familiar with the time it takes to walk from the arrival location to the interview location. Do not cut the time too short. Arrive at the interviewing location at least one-half hour prior to your appointment time.

REPORTING FOR THE INTERVIEW

You should arrive at the interview location looking and feeling your best. This means that thought should be given to the twelve-hour period just prior to the interview. Many unthinking candidates for promotional positions have worked the day before, caught an all-night fire, and found it necessary to report for the interview after only a few hours of sleep, or worse yet, none at all. While many attempts have been made to explain the slowness in response to oral boards, the excuses have usually fallen on deaf ears. It is best not to take the chance. There is too much at stake.

If you are competing for a promotional position, there has been some controversy as whether to wear a uniform or civilian clothes if given a choice. It really makes little difference. The guideline should be to dress in a manner that provides the most comfort and presents you in the best possible manner. Regardless of your choice, you should dress professionally.

Personal appearance is extremely important in an oral interview. Even if oral interviewers are instructed not to grade you on your appearance, unconsciously your appearance will have an effect on their final grade (see Figure 4-2).

Clothing should be clean and neatly pressed, and shoes should be shined to a high polish. If civilian attire is chosen, a conservative suit or sportswear with a white shirt and conservative tie should be worn. Male candidates should not wear earrings. Females should also dress conservatively in a professional business-like attire with hair style and length conforming to fire department policy. If earrings are worn, they should be conservative.

Of course, fingernails should be cleaned, and a man's hair, mustache, and beard should be in compliance with the grooming standards established for the testing jurisdiction. It is also wise that a male's haircut be at least five days old so that the hairstyle appears more

Figure 4-2 Dress neatly and conservatively.

natural. Fussy? Yes. But remember, it is foolish to give a competitor any advantage. The bottom line is that no one should present himself or herself in a manner better than you have.

WAITING

Plan to arrive at the interview location at least half an hour prior to the scheduled time. This precaution will allow time for any unforeseen incidents that might occur. Of course, arriving early means an extra wait. Waiting may add to your anxiety and, consequently, your nervousness. It is therefore best that you compensate for this.

When you arrive at the interview location, report immediately to the receptionist. The receptionist will check your name off the list, and if he or she has it, set aside your application and other material that will later be given to the oral board. If you are submitting a resume, ask the receptionist if she or he would mind giving one to each of the oral board members.

If the information is available, get the names of oral board members if you do not already know them. You should have checked with friends who have already taken the oral to see if they remembered the names of the interviewers. Alternatively, the names may be listed outside the interviewing room, or the receptionist may be able to provide them. Memorize the names prior to being called into the interviewing room. Once you are in the room, names should be associated with faces and used whenever appropriate. Appropriate means now and then, not every time you answer an interviewer's question. It is appropriate, however, to address the interviewers as Mr., Miss, Chief, sir, or ma'am.

It is best for you take something with you to read while you are waiting. Take something that can be left out of the interviewing room once you are invited to enter. A magazine, for instance, provides a variety of material that is light in nature.

It is normal to become nervous while waiting. Unfortunately, nervousness can produce a dry throat. Chewing gum while waiting will assist in keeping your throat moist. It is important to dispose of the gum prior to entering the interviewing room. A last-minute drink of water will also assist in maintaining a moist throat. This drink should be taken immediately prior to entering the interview room. A clue that it is almost time for the interview is provided when the previous candidate leaves, or when the receptionist enters the room with a new package of material. Such events usually indicate that the waiting time has been reduced to approximately five minutes.

LENGTH OF THE INTERVIEW

One of two things will generally occur when it is time for the interview. The chairperson of the board may come to the waiting room, introduce himself or herself, and then take you into the interviewing room and introduce you to the other members of the board. Or the chairperson may notify the receptionist that the board is ready, whereupon the receptionist will direct you into the interviewing room and introduce you to the chairperson. The chairperson will then introduce you to the other board members.

You should let circumstances govern in regard to shaking hands with each board member. If any board member offers a hand (as is usually the case), accept the offer. Shake hands firmly while making eye contact with the board member. Acknowledge each board member by name as eye contact is made. Remember, it is not proper for a man to offer to shake hands with a woman. He should always wait until the woman offers her hand.

This is a good place to discuss the importance of a handshake.

Tips for Success A good portion of the impression you make on each board member begins with the handshake. You should start today developing one that will give the board a good impression of your character (see Figure 4-3).

A handshake is the way most people judge another person when they first meet an individual. There is an art to the way it is done. Here are a few guidelines that might help.

Always stand and face the person you are shaking hands with. If the other person is standing, and most interviewers will be, you should offer your right hand at the navel level but not much lower. Always use your right hand, even if you are left-handed. Offer

Figure 4-3 A firm handshake while making eye contact is extremely important. A candidate should practice its development at every available opportunity.

your hand with the palm vertical. If an individual offers his or her hand with the hand down, he or she is not offering to shake hands. Don't extend your hand beyond a comfortable reach when offering it. A handshake is a method of meeting a person halfway. Therefore, do not stick yours out too far.

Make your handshake a firm grip while looking the person in the eye, but do not try to break the other person's hand. Give two or three shakes and let go. A handshake should never last more than three seconds. As previously mentioned, start practicing today. It is important that your handshake be perfected prior to the day of the interview.

Regardless of the method used to introduce you to the board members, you should enter the room in a confident manner and maintain this composure throughout the interview. Remember that your behavior will have an effect on the interviewers from the moment they come into contact with you whether or not they are supposed to be rating you.

There are some other factors regarding your demeanor during the interview that will have an effect on your grade and performance. Four of these are:

1. Remember to smile occasionally, even if you are not in the mood to do so (see Figure 4-4).
2. Make eye contact with the interviewers at least half to two-thirds of the time. More than this may give the appearance that you are coming on too strong.
3. Pay attention to your body language. Lean slightly toward an interviewer when answering his or her question.
4. Restrict your impulse to use an interviewer's name repeatedly. Once or twice during the interview is usually sufficient.

Figure 4-4 Remember to smile occasionally.

Once introductions are made, the chairperson will probably suggest that you sit down. A good procedure is not to take a seat until it is offered. If possible, move the chair to such a position that you will be able to talk to each of the board members comfortably. This means that the chair may have to be moved back three or four feet from the table. Sit in the chair in a comfortable manner that is neither stiff nor sloppy. It is best to place your hands in your lap in such a way that they can be used for emphasis but will not get in the way of your presentation.

An attempt is generally made to give each candidate an equal amount of time with the board. The allocated time will vary in accordance with the importance of the position and the number of candidates being interviewed. Most interviews last from 15 to 25 minutes. It is not unusual, however, for interviews for high-ranking positions to last much longer; interviews for entry-level positions may take less than 15 minutes.

Many candidates are disturbed because they spent less or more time in the interviewing room than was allocated to other candidates. It should be remembered that the length of time in the room has no bearing on your grade. In some cases, the board finds out all they need to know in the first five minutes. At other times, it is more difficult to draw out a candidate and it therefore takes more time to evaluate the qualities desired.

Most interviews are taped. You may be informed of this or you may not. Try not to let this bother you. The taping is done to protect both you and the board members in the event of a protest. Tapes are usually erased after their usefulness has passed.

TYPES OF BOARDS

Interviews can be handled on a one-to-one basis, or candidates may be interviewed by a board. Most fire department interviews for entrance and promotional positions use boards.

Two types of boards are in general use. For lack of better terms, they can be referred to as a **non-stress board** and a **stress board.**

The non-stress board is normally very friendly (see Figure 4-5). The board tries to place and keep the candidate at ease during the entire interview. If a candidate makes an error in judgment or a false statement, board members simply smile and continue the atmosphere of friendliness. Each member will nod his or her head in approval at the candidate's remarks, and, consequently, the candidate leaves the room feeling good. It is not until the grade arrives in the mail that a candidate questions his or her performance. The immediate reaction is generally, "How could I have gotten such a poor grade when I answered all of the questions correctly?" But, as we shall see later, answering the questions correctly, or thinking that the questions were answered correctly, is not the primary factor in the assignment of a grade. The oral board is not looking for specific answers to questions but, rather, some other intangibles.

Figure 4-5 A non-stress board is usually very friendly.

Figure 4-6 A stress board is usually more formal.

Stress boards operate differently (see Figure 4-6). A stress board attempts to evaluate a candidate's ability to operate under stress. At times, interviewers act downright mean. If a candidate uses poor judgment or makes a mistake, no matter how small, a board member may criticize him or her or increase the pressure. Board members will make candidates justify their positions and will try to intimidate candidates to make them change their minds. Many candidates leave the room sweating, wondering why they showed up for the interview in the first place. As with non-stress boards, how a person feels as he or she leaves the room is not an objective criterion for determining the grade the board will assign. Neither does the amount of sweat dripping from a candidate's brow have any bearing on the grade. Obviously, preparing for this type of board is the most painful and difficult. However, a candidate who prepares for and is capable of coping with a stress board should have little difficulty coping with a non-stress board. Consequently, this book has been designed to help you prepare for a stress board.

The interviewing board normally consists of three or more members. Representation might include the fire department, the personnel department, and some men or women from business, government, or community organizations. There will also generally be minority representation in those communities having a substantial minority population. Those chosen as board members are usually familiar with fire department operations and procedures or are experienced in hiring employees for their own organizations.

The fire department representative will usually be at least one rank higher than the position the candidate is seeking. Entry-level firefighter candidates should expect the fire department representative to hold the rank of company officer or higher. This officer is generally from the city department that is seeking the employees.

Promotional boards for company officers usually have a representative with the position of battalion (or district) chief or higher. The representative may be from the candidate's own or another department. In some areas of the country, it is quite common to use ranking officers from a nearby city that has a fire department of equal or nearly equal size.

The Civil Service Department attempts to select board members who are not likely to know any of the candidates. This choice is made for the purpose of eliminating any favoritism or bias toward any candidate. Selecting board members who do not know any of the candidates can be difficult. Consequently, board members are generally instructed to remove themselves from the interview if they feel they cannot evaluate a candidate fairly and impartially.

If a board member knows a candidate but feels he or she can still make an objective evaluation of the candidate's fitness for the position, he or she may elect to remain on the interviewing team. If you as a candidate happen to know one of the interviewers, do not try to hide the fact. However, do not try to take advantage of your relationship with the board member. If you feel you cannot receive a fair evaluation from a board member, request that the board member be removed from the interviewing team. This request, if handled properly, will not be detrimental. In fact, it may be beneficial by indicating to the other members of the board that you are not intimidated by the situation and that you have the leadership ability to take action when necessary. In this case, another person may be assigned to the board, or the board may proceed with the smaller team.

One last thought concerning board members. Regardless of the composition of the board, a thread of similarity that seems to unite members who serve is the apparent inherent desire to select the best-qualified person for the position. While board members would rather give every candidate a good grade, each seems to know the importance of his or her decisions. Without being told, they are aware of the tremendous responsibility placed on their actions. Not only are they helping to decide the future effectiveness of a department organized to save lives and property, but they also in affect are authorizing the expenditure of large sums of taxpayers' money that has been designated for the upgrading and improvement of a public service.

BRIEFING THE BOARD

Time is generally allocated to brief the board before the first candidate appears. Briefing may be from the fire department, the personnel department, or both. Whoever does the briefing acquaints board members with the general operation of the fire department, explaining the duties and responsibilities of the position to be filled. The briefer also provides some insight into what qualifications the fire department administration believes are required to do the job effectively. Those who do the briefing do not discuss any individual candidate during this period.

There are certain areas that are generally taboo during the interview process, such as sex, age, ancestry, race, religion, color, and political affiliation. Taboo areas are covered in the Equal Employment Opportunity laws and are quite technical in some respects. Many of the subjects have had to be submitted to the courts for interpretation. For example, the courts have ruled that an interviewer cannot inquire into the arrest record of a candidate. This decision was based upon the fact that a disproportionate share of minorities are arrested, and, in addition, an arrest does not mean a conviction or even that the person arrested was of bad character. Fire department applicants, however, can be asked about felony convictions because members at emergencies are placed in a position where theft is possible.

Some personnel departments provide board members with a written memorandum that helps the interviewers in their overall evaluations of candidates. Following is an example of a memorandum that could be given to board members.

> To Members of the Oral Interviewing Board
>
> As a member of the oral interviewing board, you are assuming a task that is very important to the operation of the fire department. You are asked to appraise and evaluate the training, experience, and personal traits of the candidate in regard to the requirements of the position for which he or she is applying. The objective of the interview is to provide each candidate the opportunity to prove that he or she has the training, education, and experience necessary to effectively carry out the responsibilities of the position. The burden of proof rests with the candidate; however, the evaluation of the proof is the responsibility of the interviewing board. It is important that the board help each candidate to supply the information necessary for a thorough evaluation. You should recognize and take into account that the candidate may be under extreme stress. Every attempt should be made to place candidates at ease.
>
> You have been provided information regarding the duties and responsibilities of the position for which the candidates are seeking by a supervisor from the fire department for that position. You have also had an opportunity to discuss these factors with the supervisor. Prior to interviewing the first candidate, you should discuss with the other members of the board any questions you still have. You will also have an opportunity to review the written application submitted by each candidate. This application will show the applicant's training, education, and experience. A candidate may submit a resume that will supplement the information on his or her application.
>
> There are several suitable topics that can be used for the purpose of asking questions. The following list of topics, together with a list of factors to be evaluated for each topic, has been provided to you as an aid for asking questions. Some standardization of topics and questions is necessary in any competitive examination. However, the board is given the flexibility to ask questions that seem appropriate without any regard for uniformity or follow-up. To arrive at a basis for rating the same traits in all candidates, it is often necessary to ask a variation of questions depending upon a candidate's answers and reactions.
>
> **A.** Work experience
> 1. Experience directly related to the position sought
> 2. Skills related to the job
> 3. Knowledge of the job
> 4. Deficiencies or strengths related to the position
>
> **B.** Communication
> 1. Does the candidate have the ability to follow written instructions?
> 2. Does the candidate have the ability to speak effectively?
> 3. Does the candidate have the ability to give and take directions?
> 4. Does the candidate understand the necessity for promptness and reliability?
>
> **C.** Judgment
> 1. Does the candidate have the ability to apply knowledge to specific situations?
> 2. Has the candidate had any experience in which he or she has had to use judgment in solving a problem?

3. Does the candidate have the ability to analyze situations and apply adequate judgment for determining a course of action?

D. Attitude toward the position
1. Does the candidate have a sincere desire to obtain the position?
2. Other than need, does the candidate really want this kind of job?
3. Does the candidate have any experience to back up a desire for the job?

E. Abilities directly related to the position sought
1. Will the candidate be able to work under stress?
2. Does the candidate indicate the ability to work as a member of the team?
3. Does he or she show empathy for the various types of citizens encountered in the job?
4. Does the candidate get bored easily?
5. Does the candidate have the strength and ability required for the position?
6. Will the candidate have difficulty adapting to the work schedule?

F. Stability
1. Does the candidate have a reasonable chance to remain on the job?
2. Is the candidate a happy or sad individual?
3. Does the candidate's previous work experience indicate job stability?
4. How much does the candidate know about the fire department and the community?
5. Does the candidate finish jobs that he or she has undertaken?

G. Emotional attitudes
1. What does the applicant feel is his or her weakest point?
2. What does the applicant feel is his or her strongest point?
3. Is this individual a loner or a joiner?
4. How does the candidate describe his or her temperament?

H. Overall fitness for the position
1. Does the candidate meet the quality level the fire department is seeking?
2. Will the candidate fit well into the department?

As an interviewer, you are being asked to explore all of the above points regarding a candidate's qualifications for the position. Try to avoid placing a candidate into a slot based upon what you consider a typical individual of a given race, class, occupation, or social group. Once you fall into the trap of placing an individual into a slot, you will tend to judge him or her on the basis of the characteristics you feel are typical of a person in that slot. Also avoid the **"halo" effect.** The "halo" effect is the tendency to grade an individual high on one factor based upon a favorable impression he or she has given you on a different factor.

You are being given a rating sheet for recording your evaluation of each candidate. You are encouraged to write comments on the back of the rating sheet that can be referred to for determining a candidate's final grade. Try to avoid jumping to an early conclusion. The rating sheet contains instructions for assigning points to each of the factors. This will help in determining a final grade.

It is possible that some of the candidates will be arrogant or temperamental. Under no circumstances should you permit yourself to be led into an argument with such individuals.

At the completion of the candidate's interview, you are encouraged to frankly discuss the strengths and weaknesses of the candidate with other members of the board. At the end of the open discussion, make tentative ratings on each factor together with your comments regarding that factor. When all candidates have been interviewed, each board member should rank the candidates according to the order in which he or she thinks the individuals should be hired.

Grades can then be assigned to each individual based on your ranking and tentative grades applied to each factor. Rankings and final grades should be assigned independently. The final grades of each board member will be added and an average determined. This average is the candidate's final grade.

DETERMINING A GRADE

Board members are generally briefed on the scoring process. Some jurisdictions allow board members complete freedom to assign any grade deemed applicable for a given candidate. Others afford board members considerable latitude in assigning grades but establish some grading guidelines. For example, board members may be instructed to place a candidate within a general class initially; however, board members are given the authority to assign a grade anywhere within the class. For purposes of illustration, the following classes may be established.

Superior	90–100
Above Average	80–89
Average	70–78
Not Qualified	Below 70

During the interview, board members individually place the candidate in one of the four classes. Each board member then has the freedom to establish a raw score within the range given. For example, suppose a board member deems a candidate to be "above average." He or she then establishes a grade such as 85, 86, 87, or the like, depending on how the candidate compared with others who were placed in this category. This type of system forces board members to consciously compare the qualifications of one candidate with another; however, with other systems, board members may do this, whether they realize it or not.

In some jurisdictions, board members are restricted in their grading evaluation. Rating sheets are provided and board members are instructed to assign a score for each of the qualifications to be evaluated. Scores are added to determine a candidate's final grade by a board member. With this system, each interviewer is given a score sheet that is used to evaluate a candidate on a number of different traits. The interviewer is also given a list of the factors that should be evaluated, together with questions that should be used to assist in the evaluation. The points given on each factor are added on the summary sheet to determine a candidate's overall rating.

Still other jurisdictions use a combination of the two systems. Graders are requested to indicate a mark for the candidate in various traits; however, the final grade is left to the discretion of the grader, using his or her marks on the various traits as guidelines (see Figure 4-7).

INTERVIEW SCORE SHEET

1-not acceptable 2-below average 3-average
4-above average 5-superior

1. EDUCATION AND TRAINING	1	2	3	4	5
2. FIREFIGHTING EXPERIENCE	1	2	3	4	5
3. KNOWLEDGE OF THE POSITION	1	2	3	4	5
4. VERBAL AND COMMUNICATION SKILLS	1	2	3	4	5
5. PERSONAL PRESENTATION	1	2	3	4	5
6. POISE AND CONFIDENCE	1	2	3	4	5

TOTAL ________

RATING SCALE

Excellent	(26+)	________
Above average	(22-25)	________
Average	(17-21)	________
Below average	(12-16)	________
Not acceptable	(11 and below)	________

INTERVIEWER RATING ________

COMMENTS

CANDIDATES NAME ________________________________

BOARD RATING ________

INTERVIEWER'S SIGNATURE ________________________________

Figure 4-7 A sample interview score sheet.

WHAT THE BOARD KNOWS ABOUT CANDIDATES

Generally, before a candidate enters the boardroom, time is allotted for board members to review whatever information is available about the candidate. The amount of material to which they have access varies from jurisdiction to jurisdiction and from position to position. More information is usually available for promotional positions than for entry-level positions. The minimum amount of information available to the board is what appears on a candidate's application.

If a candidate submits a resume with the application, the resume is also normally available to the board. Applications and resumes were discussed in Chapter 3.

Additional information for promotional positions may come from a candidate's personnel file; however, not many jurisdictions submit this information to the board. If used, the information might include department ratings, sick record, work assignment record, complaints filed against a member, awards, commendations, and similar material that will assist the board in arriving at an objective conclusion regarding a candidate's suitability for the position. All the available material may be discussed openly among the board members, or each member may review the material individually and without comment. Nevertheless, most board members make a few notes of items they wish to discuss later with the candidate.

One item interviewers usually do not have is the result of a candidate's performance on other portions of the examination process. This information should have no influence on the oral interview grade.

LINE OF QUESTIONING

The line of questioning can take any tack. However, opening questions are usually designed to relax a candidate and get him or her to talk. This technique is true regardless of the type of board formed. Consequently, a candidate can expect opening questions about his or her family, work history, hobbies, and so forth. These are subjects that are thoroughly familiar to a candidate. Therefore, he or she should be able to discuss them in a knowledgeable and confident manner.

As one of the objectives of opening questions is to get the candidate to talk, questions are seldom worded in such a way that a simple "yes" or "no" answer would suffice. Following are some samples of opening questions for the position of firefighter. Each of these questions should be written on a 3×5-inch card for use in the practice sessions. The completed cards should be placed in your card file with others you have prepared.

- Tell us something about yourself.
- What makes you think you are qualified to be a firefighter?
- What qualifications are required to be a firefighter?
- Tell us what a firefighter does.
- What are some of your hobbies?
- What is your educational goal?

- What do you do in your spare time?
- What are some of your interests, other than the fire department?
- Do you have any relatives or friends who are working for the fire department?
- What do your primary friends think about you becoming a firefighter?
- In what sports did you participate in high school?
- What type of community activities have you been involved in?

Opening questions for promotional positions may follow the same line but will usually consist of one of the following questions. In fact, one or all of these questions will normally be asked early in the interview regardless of the position applied for. The thought process for answering these questions should be developed prior to preparing answers to the questions. More information regarding the development of the thought process will be provided in Chapters 5 and 6. Each of the questions should be written on a 3 × 5-inch card for use in the practice sessions. Store the completed cards in your card file together with the others you have prepared.

- What makes you think you are qualified to be a company officer?
- Tell us why you want to be a company officer.
- Tell us exactly what a company officer does.

After the opening questions, the board may ask a few more general questions, or the interviewers may delve into something that caught their attention on the application or resume. Sometimes they will jump immediately into problem-solving situations. Problem-solving questions may involve either personnel or fire-fighting situations. They are designed to determine a candidate's thought processes, common sense, and communication ability. Guidelines for answering problem-solving questions will be offered in a later chapter. This line of questioning will generally continue until the allotted time is over or until the board has found out what they want to know about the candidate.

This process is usually used providing the board has the authority and flexibility to ask questions based upon the responses of a candidate. However, some jurisdictions require that the identical questions be asked of every candidate for the position of entry-level firefighter. This process is implemented when a jurisdiction is leery of being sued as a result of a candidate's claim that he or she has been discriminated against based upon sex or race. Others follow this line of thought primarily because of federal guidelines.

TERMINATION OF THE INTERVIEW

When the allotted time is nearly up, the chairperson will generally ask a candidate if there is anything he or she would like to add. The purpose of this question is to ensure that every candidate has had the opportunity to tell the board everything he or she wanted to talk about. It is not an indication that something has been left out or that the board wants to hear more. In fact, the opposite is generally true. What the chairperson is really saying is, "We have found out what we wanted to know about you and have determined the grade we intend to give you. If you want to try to change our minds, go ahead, but you don't stand much of a chance." In general, it is wise for a candidate to reply that everything has been well covered, thank the board, and leave. However, if something extremely important has been left out, a candidate may want to take time to cover it. In most cases, any further statement can do more harm than good.

POST-INTERVIEW PROCEDURES

After the candidate has left the room, the board members will assign an individual grade to the candidate's performance. Occasionally, the candidate is discussed openly. However, this is usually done after each board member has assigned a grade, or the instructions given to the board included that an open discussion should take place prior to the assignment of a final grade. If an open discussion does take place, it is rare for a board member to make a drastic change in a grade as a result of this discussion.

RECEIVING NOTICE

The results of the interview are usually mailed within two weeks after the last candidate has been interviewed. The length of time required for all candidates to be interviewed varies according to the number of candidates and the number of boards established for interviewing. The total length of the interview process is closely related to the size of the community or examining jurisdiction. Some interview periods are completed within a week, while others extend over several months. For example, the interviewing process for the position of police sergeant for some larger cities has, at times, extended over a four-month period.

THE PROTEST PERIOD

A candidate's score on the oral interview is determined by the candidate's performance in the interviewing room. However, many candidates feel they should have received a much higher score. Few believe that their score should have been lower. Most jurisdictions do not consider a difference of opinion between board members and candidates regarding a score as a valid reason for protest. However, most do provide some means for candidates to express unhappiness if they feel they have been discriminated against because of prejudice or if a candidate believes the examination was conducted in a fraudulent manner. One of the objectives of the Civil Service Commission is to ensure that examinations are conducted fairly, impartially, and in accordance with the rules and regulations they have established. Consequently, most have made arrangements for a candidate to protest the results of an examination if he or she feels that the rules and regulations have been violated. Candidates should be aware that most regulations establish a time limit and a method by which protests must be filed. Generally, all protests must be in writing. They should be substantiated with facts, not opinions.

Unfortunately, most candidates who fail examinations attempt to blame some other person for their failure. Those who honestly evaluate their background, experience, and performance in the interview normally realize that personal improvements need to be made. Unless there is definite proof of prejudice or fraud, a candidate is wise to accept the grade allocated and do as much as possible to improve himself or herself for the next examination.

PRE-APPOINTMENT INTERVIEW

If an entry-level candidate is successful in obtaining a high enough place on the Eligible List to be considered for hiring, then he or she should expect that the interviewing process has not ended. Remember that under the Rule of Three, two more candidates than the number of openings that exist are high enough on the list to be considered for hiring. Entry-level candidates in jurisdictions that have only a few openings will generally be required to be interviewed by the fire chief before being appointed. The chief may conduct the interviews himself or herself, or the chief may invite one or more his or her command officers to assist in the process. In large jurisdictions that have a great number of openings, the interviewing may even be done by staff officers, or perhaps not at all.

Prior to the interview, the chief will obtain from the Civil Service Department as much information as possible about a candidate. This material will generally include the application, resume, results of the background check, and any other information the Civil Service Department or the fire department staff has gathered that will assist the chief in making a selection. The grades given by the individual members of the interviewing board are normally not provided.

The interview is usually conducted on an informal, non-stress basis. The objective of the interview is to allow the chief to learn as much as possible about those who are being considered for hiring. The discussion is generally limited to a candidate's background, family, hobbies, and so on. Although this interview should not be taken lightly, candidates who have successfully passed the regular board should not fear it.

Some chiefs in large jurisdictions also use the pre-appointment interview for promotional positions, particularly for appointments at the chief officer level. The reason for this is the major effect that these officers will have on seeing that department policies and procedures are carried out. Without the pre-appointment interview, it is possible in a large jurisdiction for an individual to place high on the Eligible List for a high-ranking position without the chief really knowing too much about that person.

SUMMARY

This chapter was primarily an informational chapter. In it, you were provided information regarding the overall interview process, the types of interview boards you might encounter, information and instructions that might be given to the interview board, an example of a rating sheet that could be used, the line of questioning you might expect, and a suggestion on how to terminate the interview. With the exception of the instructions to make more 3×5-inch cards, very little participation material was included in the chapter. However, the information that was included has moved you a little further down the path you are traveling.

REVIEW QUESTIONS

1. What is being determined during an oral interview?
2. What is one of the primary reasons that many people are discharged from a job?
3. List ten personal traits that the oral board is attempting to measure.
4. What is the first positive indication you will have that you are going to participate in the oral interview portion of the examination process?
5. What is important about the twelve-hour period immediately prior to the interview?
6. When should you plan on arriving at the interview location?
7. What can you do to help keep your nervousness from producing a dry throat prior to the interview?
8. What are the guidelines for shaking hands with board members?
9. What should you remember about shaking hands with a woman?
10. Approximately how long do most interviews last?
11. What are the two types of interviewing boards?
12. In preparing for an oral interview, it is best to prepare for what type of board?
13. How many people are normally on a fire department oral interview board?
14. What should you do if you know one of the members of the interview board?
15. What are some of the subject areas that are generally taboo during the interview?
16. Can an interviewer for a fire department examination inquire into the felony conviction of a candidate for a crime? Why?
17. Is a board member always given complete freedom concerning the score he or she wishes to assign to a candidate?
18. Can an interviewer change the grade he or she gave a candidate once it is written down?
19. A candidate received the following grades from three interviewers: 88, 92, 93. What is his or her final grade on the interview?
20. What material about the candidate does the oral board have to review prior to the candidate entering the room?
21. If the board has the personnel file on a member who is seeking promotion, what information will board members probably have access to?
22. What are the opening questions generally designed to do?
23. Regardless of the position applied for, what two questions will probably be asked early in the interview?
24. What are problem-solving questions designed to determine?
25. When the time is nearly up, the chairperson will generally ask if there is anything else the candidate would like to say. What is the purpose of this question? How should you normally respond to it?

ANSWERING QUESTIONS

LEARNING OBJECTIVES

The objective of this chapter is to acquaint the reader with the general procedure used during an interview session and the principles of answering questions. As the result of reading this chapter, the reader should be able to:

- Describe what happens prior to the questioning.
- Explain what type of questions will most likely be asked during the opening moments.
- Explain the two most common errors made by candidates in answering questions.
- List the dos and don'ts that should be considered by a candidate during the interview session.

INTRODUCTION

Contrary to what many candidates who have been subjected to their inquisition think, oral board members are human. Most would like to see you do your best. Consequently, oral board members will generally try to assist you in putting forth your best effort. Their job will be difficult, however, if you withdraw into a shell and fail to display your real personality.

One of the first offers the board chairperson will most likely make to you is to have a seat and make yourself comfortable. A comfortable position during the oral interview should be one where you are neither sitting rigidly nor sloppily. Some candidates will find hands to be a problem. One method of controlling your hands is to place them in your lap so that they can be used for emphasis, if desired. However, they should not be in a position in which you are tempted to use them to wipe your nose, flick your ears, or wipe the sweat off your forehead.

Much of the impression you make on board members will depend upon the way you conduct yourself. It is important that you be courteous, alert, and self-confident at all times.

Smile occasionally and speak in a self-assured tone of voice. Sitting in a comfortable position and maintaining control of your hands will assist in providing an impression of confidence.

Tips for Success Maintain your professionalism at all times. It is important that you project an aura of enthusiasm, confidence, and ambition.

Company officer candidates should be alert for any opportunity to praise their department or top managers on the manner in which the department is operated. If the opportunity presents itself, entry-level candidates should also praise the department, saying they have heard from friends that it is a good organization. Show pride in the department's achievements whenever possible.

Learn to talk *with* board members, not at them. Make an effort to talk freely, maintaining a social atmosphere rather than a formal one. This behavior provides board members with a better opportunity to gain a true understanding of your personality and some idea as to how you will probably operate with your peers, subordinates, and superiors.

When you talk with board members, it is extremely important that you make eye contact. Personal rapport should be established with each member. If you look at the table, the ceiling, out the window, or across the room, you will not make your personality felt. Every time you make personal contact with a board member through good eye contact, you sell a little more of yourself. If a board member happens to be looking down or perhaps taking a drink of coffee when you attempt to make eye contact, then skip to another member, but come back to the previous member at another time. It is important to make contact with each board member as often as possible.

FIELDING QUESTIONS

You will be asked a number of questions during the interview. Some will be phrased simply to get you to talk. Others will be asked so that board members can gain a better understanding of your background and qualifications. Still others will be directed to gain a better understanding of your comprehension, alertness, judgment, and thought process.

Tips for Success It is important for you to realize that there are no right or wrong answers to most of the questions. Questions are not designed to test your technical knowledge. This testing was done on the written examination. The board is, however, looking for the answer to a question that will not be asked: "How would I like to have this candidate work for me?"

If you get a positive "yes" on this question, you will get a good grade. On the other hand, an answer of "no" will also be reflected in your grade.

Although there are no right or wrong answers to most questions, there are some general errors made by most candidates when formulating responses. Two of the most common are:

1. A candidate will try to answer a question the way he or she thinks the board wants it answered.
2. A candidate will fail to answer the question.

You will make a mistake if you answer a question the way you think the board wants it answered. Questions will be phrased to determine your thought process. Board members want to know what you would really do in the circumstances presented and why you would take that particular action. You should be honest with the board and yourself, and not try to play games.

Failure to answer a question is usually the result of poor organization. Most candidates talk around the answer with the intention of summarizing at the end. However, they normally merely ramble and, in fact, fail to answer the question. The best method of avoiding this is to reverse the usual organization. Get into the habit of saying, "Here is what I would do" and, "This is why I would do it that way."

In addition to avoiding the two common errors just discussed, be aware of the dos and don'ts discussed in the following section.

DOS AND DON'TS

Dos

- *Be comfortable.* Sit erect but not stiff. Use your hands for emphasis, if you wish. Make sure your hands do not become a distraction by drumming on the table, wiping your brow, and so forth.
- *Control the interview.* Controlling the interview does not mean that you will be able to decide what questions are asked. That is impossible. Control means that you tell the board only what you want them to know. You exercise control by simply answering the question, and not by blurting out additional information because the board members sit there and stare at you. When you have answered the question, sit and wait for them to ask the next one.
- *Be confident but not cocky.* You should leave the board with the impression that you can do the job in a confident and efficient manner. Cockiness generally leaves a negative impression, and many times is viewed as a mannerism to conceal some weakness. The line between confidence and cockiness is very thin and extremely difficult to define. However, the difference is recognized by board members who have had limited interview experience as well as those whose interviewing experience is extensive.
- *Be friendly and courteous, and talk with board members, not at them.* Talk to the board as you would to a friend. However, it is not degrading to you if you use "yes, sir" or "no, ma'am" when answering questions. Give board members the respect due

them, and never forget that the impression you are making on them is your pathway to success or failure.

- *Be attentive and understand the question.* Give each board member your undivided attention. Make sure you thoroughly understand each question. Do not hesitate to ask for more information if it is needed. Some board members will deliberately withhold information to see if you will take action without all the facts. Direct the answer to the member who asked the question; however, make eye contact with other members of the board when answering.
- *Answer questions promptly, but do not be too hasty.* If you know the answer to a direct question, answer it promptly. If the question takes some thought, make sure you consider all options before answering. It is impossible to retract a statement, once made.
- *Admit errors.* If you make a mistake, and the mistake is pointed out by a board member, do not hesitate to say, "I made a mistake." It is not wise to hold stubbornly to a position that has proven to be wrong. On the other hand, it is generally better to stand firm on a position taken if there are several possible solutions and you truly believe your plan of action is best under the circumstances. Many board members will deliberately challenge a candidate just to see if he or she is willing to defend a position. Although it is not wise to debate with a board member, it is best not to surrender a good position. If a position is worth taking, it is worth defending.
- *Use common sense.* Over a hundred years ago, Ralph Waldo Emerson wrote, "Nothing astonishes men so much as common sense." Converted into a principle for answering questions during an interview, it can be stated as

Tips for Success When all else fails, use common sense.

This principle should be applied throughout the interview. Of all the dos and don'ts offered, the application of this principle is no doubt the most valuable tool available for you to solve problems.

Don'ts

- *Don't argue with a board member.* Although it might be necessary to defend a position, do not debate or argue the point. It is possible that you could win the battle and lose the war. Remember the saying, "He was right, dead right as he rolled along, but he's just as dead as if he were wrong."
- *Don't interrupt.* Make sure a board member has finished asking the question before you attempt to answer it. Many candidates are so anxious to show they know the answer to a question that they start talking before the board member has a chance to finish asking the question. Developing the habit of allowing a slight pause between the question and your answer will usually prevent this mistake. If necessary, get

in the habit of counting silently to three to yourself after the board member has finished asking a question.

- *Don't be intimidated.* With a stress board, questions could be shot at you fast and furiously. Do not let this technique alarm or disturb you. Answer each question separately. It may be possible that board member B will shoot a question at you while you are answering a question from member A. You may hear the second question and you may not. If you do hear it, wait until you finish your response to A's question, then turn to B, repeat his or her question, and answer it. If you did not hear the question, then apologize to B for not hearing it and ask if he or she would mind repeating it.
- *Don't exaggerate.* State facts and be honest when explaining your background and experiences. Remember that in today's world, it is very easy to verify everything you say. It is much better to be humble than to oversell.
- *Don't wisecrack.* The time spent in the interview room is extremely important. Furthermore, it is a very serious situation. While every attempt should be made to create and develop a friendly thought exchange, the time available should be used to best advantage. Keep poised and professional at all times
- *Don't become negative.* Maintain a positive attitude during the entire interviewing process. A board member may try to get you to think negatively. Make every effort in such situations to change negative situations to positive situations.

A board member may ask, "What is the biggest mistake you've made since you've been on the job? Without a doubt, every candidate has made mistakes, and this type of question could cause a candidate to start thinking of each one, try to pick out the biggest, and then tell the board about it.

Tips for Success Remember, mistakes that are properly handled can become assets.

For example, many firefighters spend years playing rather than studying for promotion. When they recognize their error, they put their nose to the grindstone and try to make up for the wasted years. Such a change of mind results in an error being turned into an asset. A negative situation has been turned into a positive one.

When fielding this type of question, remember that traits you might think of as weaknesses might be considered strengths by the board. Two examples of factors that fall into this category are perfectionism and impatience in getting the job done.

Some questions require a negative reply. Answering these questions does not indicate that you have become negative. For example, suppose the board asks you to describe the least efficient officer you have ever worked for. Notice that they have given you no alternative. You cannot answer that you have never worked for an inefficient officer because they are not indicating that any of the officers you worked for were inefficient. They are merely asking that you describe the least efficient officer. Go ahead and describe him or her.

However, it would be an error to divulge his or her name. If you are asked to do so, reply that since you are being critical, it would be best that you not do so.

- *Don't air your dirty linen.* Problems within an organization should be kept within the organization. They should not be aired for the whole world to see. Every candidate has worked for some supervisor he or she thought was not as efficient as he or she could have been—and every candidate has worked for an organization that needed some drastic changes. Do not voluntarily mention these things. Show loyalty to your organization and for those you previously worked for, knowing that improvements are always possible. Board members sense that if you are not loyal to your current organization or superior, then you probably will not be loyal to others in the future.
- *Don't ramble.* Provide sufficient information to fully answer a question, then stop. Time is at a premium. Give board members time to ask questions rather than use up time with extraneous comments.
- *Don't guess.* If you do not know the answer to a question, say so. There is nothing wrong with saying, "I don't know." It is much better to say this than to try to bluff your way through a situation.
- *Don't dwell on your current job.* Your current position is important only with respect to how it relates to the job you are seeking. Tie in what you have learned in your current job that helps qualify you for the new position, but do not try to impress the oral board with how important your current job is and how well you do it. If the job is so important, and you are doing it so well, the board may decide that it would be better for you to keep it. More than one board member has said, "Well, he's convinced me that he's a good apparatus operator, but I'm not sure whether or not he's qualified to be a company officer."
- *Don't use abbreviations or technical terms.* Some of the board members may not be fully acquainted with all the facets of the position being sought or with your current occupation. For example, use the full name National Fire Protection Association rather than the initials NFPA. If it is necessary to use a technical term, then it is wise to explain the meaning to those who may not be familiar with it.
- *Don't dominate the interview.* Through the application and resume, you might try to guide the board into asking questions you would like to discuss. However, you should not try to dominate the interview by talking too much. There are certain things board members want to find out about you. Give them a chance. If you spend all the time talking, it gives board members little opportunity to ask questions. Answer the question. Stop talking. Wait for the next question.

SUMMARY

This chapter was primarily an informational chapter. It was designed to acquaint you with some of the dos and don'ts of answering questions. Your participation was one of adopting the principles for answering questions as presented and outlined in the chapter. As you continue down your the chosen path, the one principle to keep in mind is, "When all else fails, use common sense."

REVIEW QUESTIONS

1. What is one of the first offers the board chairperson will probably make to you?
2. What is considered a comfortable position for sitting?
3. What is a good method to ensure that your hands do not get in the way during the interview?
4. What should you remember about eye contact?
5. What should you do if you try to make eye contact with a board member and he or she is looking down?
6. What are the two most common mistakes made by candidates in answering questions?
7. Why should you avoid answering a question the way you think the board wants it answered?
8. List the two things the board members want to know when they present you with a problem.
9. What is usually the cause of a candidate failing to answer a question?
10. What should you get into the habit of doing to avoid the trap of not answering a question?
11. What is meant by the phrase "controlling the interview"?
12. What should you do if you do not fully understand the question asked?
13. If you know the answer to a direct question, how fast should you answer it?
14. What should you remember about answering a question that takes some thought?
15. If you make a mistake, and the mistake is pointed out by a board member, should you admit the mistake?
16. If you take a stand on a position to which there are several possible solutions and a board member challenges you on your position, what should you do?
17. What is a good principle to remember about common sense?
18. Where would the principle of "You could win the battle but lose the war" most probably apply?
19. What should you do if you are answering a question given to you by one board member and another board member shoots a different question at you?
20. What type of response should you try to make if a board member tries to get you to become negative?
21. How can you avoid rambling?
22. What should you do if you are asked a question to which you should know the answer but do not?
23. What should you remember about the use of abbreviations and technical terms?
24. A good procedure to use during an interview is not to dominate the interview. How can you avoid this error?
25. What kind of questions are asked to test your technical knowledge?

CHAPTER 6

OPENING QUESTIONS

LEARNING OBJECTIVES

The objective of this chapter is to introduce the reader to a process to follow to answer two questions that will most likely be asked in the early moments of an oral interview. As a result of reading this chapter and using the procedures introduced, the reader should be able to prepare his or her answers to the following questions:

- Why do you want the position?
- What makes you think you are qualified for the position?

INTRODUCTION

The importance of using common sense will be stressed throughout this book. Yet, it is discouraging to see how often this principle is violated. Imagine a class in basic firefighting principles. During this class, the students are told that a particular question will be asked on the final examination. Common sense dictates that *every* student should answer the question correctly on the final examination. If not, then the only logical explanation is that some students are just not motivated enough to indicate the question in their notes and learn its answer. It is logical to assume *you* are motivated enough, or you would not be reading this book.

You will be given two questions, of which at least one will be asked during your oral interview. The chance is good that both will be asked during the interview for entry-level firefighters. Moreover, one of these questions will probably be one of the first asked. But there is something even more important. *Most authorities on oral interviews will tell you that the grade you receive will be determined within the first five minutes of your interview, and, in many cases, within the first two minutes.* Therefore, it does not require much common sense for you to realize that a good portion of your grade may be determined by the way you respond to one or both of these questions. Furthermore, anyone who reports for an oral interview who has not given some thought to and prepared an outline for a response to these two questions is just not using common sense. These two questions are:

1. What makes you think you are qualified to be a firefighter? A company officer?
2. Why do you want to be a firefighter? A company officer?

These two questions may not be asked exactly as shown. For example, question 1 may be phrased as: "What qualifications do you believe are required to effectively assume the position of firefighter?" or, "Why do you believe you are qualified to be a company officer?"

Regardless of how the question is asked, the board is really asking you to explain to them what you consider to be the qualifications for the position and why you think you have them.

WHY DO YOU WANT TO BE A FIREFIGHTER? A COMPANY OFFICER?

Surprisingly, this is not an easy question to answer. The first thought that might jump into your mind is money. However, this is seldom the primary motivating factor for a firefighter wanting to become a company officer. In fact, for promotional positions, money is rather far down the scale. Candidates have been asked during interviews if they would accept the company officer position if there were no increase in pay. Most responded that they would. It is possible they were making one of the primary errors made by individuals in answering questions asked by the board—answering a question the way you think the board wants it answered. However, in this situation, it is unlikely that this is the case. Statistics have shown that people will accept a change from a blue-collar position to a management position with a decrease in pay if there is an opportunity to increase their salary in the future.

When you consider this question, take a look at the firefighter's job first. Candidates for this position are generally young men and women who are competing for a position in an organization in which they will probably spend the rest of their working days. Ideally, what is it that an individual wants when selecting a lifetime career?

A career might be thought of as a balancing force between job satisfaction, economic needs, and security. Of course, each individual is different, but the following list probably contains most of the variables an individual wants in a job. The variables are listed in alphabetical order as the priority will change from individual to individual.

- *Adequate pay.* Note the word "adequate." Early in their careers, most people would like to earn enough to provide their families with the basic necessities of life and have a little extra money to use as they wish. They expect their pay will increase in time, even without an advancement.
- *Challenge.* Most people state they want a job which involves work of such a nature that they will have to use their mental and physical abilities to perform it.

- *Chance for advancement.* It seems that there is a human desire to improve one's position in life. Most people want a job that offers the chance to advance, if they so desire.
- *Excitement.* A stirring of your innermost feelings. A pounding of your heart with an increase in your blood pressure. People have been seeking excitement since the beginning of time. The ultimate satisfaction is to have a job that provides it.
- *Fringe benefits.* These benefits include the basic day-to-day needs of workers in the areas of health care coverage, dental care coverage, vision care coverage, and life insurance coverage. In many work environments, fringe benefits are the primary factors in workers' negotiations with management.
- *Interesting.* The reason we continue to work at a particular job is usually because it's interesting.
- *Opportunity for variety.* Some people are capable of turning the same bolt day after day, week after week, and year after year, but most are not. Most people prefer a job that is different or somewhat different every day. This factor takes the drudgery out of a job, and makes it interesting.
- *Opportunity to help people.* Although some individuals do not show it, most like to help others. Most people respond quickly to other people's needs. The ultimate occurs when the help you provide results in the saving of someone's life.
- *Pension.* People seeking a career realize they will not work all their lives. Someday they will have to retire. Financial security for themselves and their families during this later period of life is extremely important in the selection of a career. Some organizations, such as the military service, use this factor as a prime recruiting tool.
- *Pride.* Most people want a job of which they can be proud. They want to belong to an organization that has respect in the community. They want to be proud when they say, "I am a firefighter."
- *Security.* This means job security—a job that will be there regardless of a depression, a national emergency, or a change of government. People want to live in an atmosphere of security, knowing there will always be a paycheck for them to pay their bills.
- *Time off.* They want time to spend with the family, to pursue an education, and to engage in hobbies. People want to work; however, they prefer a job that provides them with the free time that makes life worthwhile.
- *Vacation.* Most jobs provide vacation time with pay. The time allocated for a vacation usually increases with seniority. In addition to the amount of vacation provided by a job, the freedom to take it when one wants it is important.
- *Work with people.* Some individuals are content to remain in a place where they can work by themselves. Most people are not. The majority of people want a job where they work with others. If no one else is around, the job may become boring, dull, and uninteresting.

There may be other factors, but the preceding list contains many of the reasons people want to work. Now, how do you use these in an oral interview? Let's take a look at a candidate applying for the position of firefighter who is asked, "Why do you want to be a firefighter?"

Although there are many possible responses to the question, the answer may go somewhat like this:

> I've reached the point in life where I think it's extremely important to select the career in which I intend to spend the rest of my working life. This is an important decision. A few months ago I sat down and tried to analyze exactly what I wanted in a job. Some of the things that seemed important were a job that would provide a challenge, one that would offer a chance for advancement, one in which I would work with people and have the opportunity to help people, and one that would provide enough variety and interest so that I would look forward to going to work. One of the important factors for me is a job that will provide a sufficient salary to take care of the basic needs of me and my family, and one that will offer good fringe benefits such as a good health and dental plan, an adequate vacation, and time off to pursue some of the things I want to do. I also considered that it was important for the career to have job security and an adequate pension system.
>
> In addition to looking at these things, I also took a good look at the fire service. I found that the fire service offers everything I want in a job.

This answer should provide you with some idea as to how to go about formulating a response. Contrary to some opinions, you should include pay and fringe benefits in your answer. However, mention them near the end of your answer. Anyone who says these things are not important considerations in a job is not being realistic or truthful.

Another thing that should be noted in the presentation is that the candidate opened a door when he or she said he or she wanted time off to pursue some of the things he or she wanted to do. It should not surprise the candidate if the next question is "What are some of the things you want to do with your time off?" Of course, if the candidate comes back with a response such as, "I would like to further my education so I'll be better prepared for a promotion when the time arrives," it would probably be a help; but remember, some answers could also be detrimental. The principle that is proposed is this:

Tips for Success During your interview, be careful about opening doors to other subject matter. Open those you wish opened, but keep the others closed. When you open a door, be prepared to respond positively to the question that may follow.

Of necessity, the answer to the question of why you want the job asked at the promotional level would be quite different from that at the entrance level. A candidate applying for the position of company officer would already have all the things an entrance firefighter is looking for. It is, therefore, necessary to analyze why a person wants to be promoted to a position where there are more responsibilities and possibly more headaches.

Some people will say that wanting to advance is a basic human desire. This is not necessarily true. There are many good firefighters who are satisfied with their positions in life and do not want the responsibilities and headaches of an officer's position. If you are seeking the position, have you really sat down and analyzed the position and then asked yourself why you wanted it? Take a look at some of the reasons individuals want to enter the management arena. Perhaps some of these reasons will apply to you.

Again, the factors will be listed in alphabetical order because the priority will vary from individual to individual.

- *Challenge.* Below an officer's position, work is accomplished primarily through the use of a firefighter's hands. At the company and higher levels, work is accomplished through the use of a different tool—people. The challenge is to be able to accomplish the job by using this new tool. Of course, there is also a mental challenge in the planning, organizing, coordinating, directing, and controlling of the operations.
- *Chance to participate in organizational decisions.* In some departments, a firefighter has some degree of input into the practices and procedures of company operations. However, his or her voice is seldom heard above the company level. If a person is proud to be a member of an organization, he or she is most likely to appreciate having some say in the direction the organization is headed. The higher you move in an organization, the more chance you have of influencing the department's direction.
- *Chance to put into effect some of the ideas you have regarding company operations.* Most people have various ideas as to how they would do the job better if they were in charge. Sometimes these ideas change into a burning desire to see them implemented. Whether or not the ideas will work, the chance to put them into effect is an excellent motivating factor.
- *Increase in standard of living.* It takes very little effort for most people to live up to the limit of their income. There seems to always be a need for just a little more money. The chance to receive it and therefore increase one's standard of living is a strong motivating factor for many people.
- *Increased pension and pay benefits.* The higher you move in an organization, the more pay you receive. More pay leads to a larger pension. Many pension systems are based on a percentage of the average of the last three years' or the last years' pay. Planning for retirement seems to increase in importance as an individual approaches the point at which he or she could retire if he or she wishes to do so.
- *More control over the working environment.* The individual who plans the day's activities has much more say over what will be done than those subject to a superior's decisions. A person in the position to make the decisions has much more control over his or her life. This is extremely important to some people.
- *Prestige.* This is a factor that some people hate to admit is important; however, it provides more job satisfaction to those who need it.
- *Sense of accomplishment.* The feeling that all the effort that has been put into preparing for promotion, both on and off the job, has been worthwhile.
- *The opportunity to use one's education and experience.* Most individuals prefer an opportunity to share their education and experience with others. Education and experience neither shared nor used for the benefit of the organization are wasted. Being in the position to use his or her education and experience provides an individual with more of an opportunity to contribute to the organization.

(Note: The first level of company officer in some departments is captain. In others it is lieutenant.)

Using the preceding ideas, the answer to "Why do you want to be a company officer?" may go something like this:

> There are many reasons why I want to be a company officer. The most important is that it will give me the opportunity to use the education and experience I have that qualifies me for the position. A second reason is that it will give me more input into the direction the department will move in the next few years. I am very proud of this department and would like to do whatever I can to see that it continues to progress. The third reason I want the position is that it will provide me with an opportunity to put into effect some ideas I have for improving company operations. Of course, I would be remiss if I didn't mention that it will also increase the standard of living for both myself and my family.

Note again that this presentation has opened a couple of doors. The most notable one would probably be followed by the question, "What are some of the ideas you have for improving company operations?" Be sure that when you open a door and give interviewers an opportunity to obtain more information from you that you are prepared with a good response.

There are several responses to the question of why you want to be a company officer. It is imperative that you analyze why you really want the job and prepare a logical response to the question. Furthermore, be careful about opening doors. Don't hesitate to do so when you would like to and are prepared to talk about the subject. Opening doors can be a very effective tool for making a good presentation.

WHAT MAKES YOU THINK YOU ARE QUALIFIED TO BE A COMPANY OFFICER?

There is a chance that you are not fully qualified for the position. This statement is in no way intended as a reflection on your ability, but more as a statement of fact regarding the system used for hiring. Unfortunately, fire departments are one of the few types of organizations that hire people who have the potential for doing the job rather than hiring those who are fully qualified and have had experience in the position they are seeking. This fact is true at both the entrance and promotional levels.

Before the appointing authority of an organization hires a typist, a test is given to the person to make sure he or she knows how to type. An accountant will be thoroughly tested prior to being hired to make sure he or she knows how to maintain the books. But a fire chief may hire a firefighter who does not know how to lay hose, raise ladders, or provide emergency care. Many departments have eliminated this deficiency by establishing the rule that a candidate for the position of firefighter must be Firefighter II and EMT certified in order to take the examination. Although such a requirement eliminates a portion of the factors necessary for an individual to operate efficiently as a firefighter, important intangibles still exist, such as the ability to get along with people, the ability to present himself or herself effectively, and the all-important factor of being able to operate as a member of the team.

The fire chief will also promote a firefighter to captain, pin a badge on him or her, and in an unvoiced statement declare, "Yesterday you were a nobody, and today you know all the answers." And it is possible that the man or woman chosen has not been in charge of a company for a single day and, furthermore, has not been given a single day of department training on how to operate a company. Despite this fact, some of those who have badges fastened to their uniforms have been officers from the first day they reported to their new assignment. Unfortunately, there are others who never truly fill the shoes.

The fact that fire departments hire potential achievers rather than fully qualified individuals has not been mentioned in a derogatory manner, but rather to help you in your preparation for your interview. Many candidates will develop a negative attitude when they analyze their backgrounds and experience and find there are gaps in regard to their so-called ideal qualifications for the position.

Tips for Success Keep in mind that the oral board is evaluating your *potential* for the position, rather than your background against an ideal.

In addition, a change has been made in the fire service regarding the hiring of potential individuals for positions rather than fully qualified people in many parts of the country. A number of community colleges are offering fire academy training for pre-service students and courses in management, computer training, and supervision for those desiring to be officers. The students graduating from fire academies know how to lay hose lines, raise ladders, throw salvage covers, and so forth. Some academies even certify students as EMTs. The courses offered for potential officers help eliminate the deficiency that a badge can be pinned on an officer who has had no training for the position. It is wise for people who live in areas where this training is offered to do their best to enroll in these courses. Those who do not give the competitors an advantage in the examination process. Remember, it is important to have all the background and experience your competition has, plus a little bit more.

Now, what approach should be taken when you are formulating an answer to the question, "What makes you think you are qualified to be a company officer?" First of all, you should recognize that the question may not be directed at you in exactly that form. Instead, interviewers may state, "Tell us about yourself," or something similar. Be alert to the fact that the board has opened a door rather than you having to open it. Of course, you want to explain to the board why you are qualified for the position. Be prepared when you are telling the board about yourself that you include in your discussion why you are qualified for the position. If you fail to do so, you may pass up the only opportunity you have of giving your sales pitch. For example, if you are asked "Tell us about yourself," you might ask the board if it is all right to confine the discussion about yourself to your qualifications for the position, or you might give them a quick sketch about your background and then ask if it is all right to relate this background to your qualifications. You will find that the board will normally be more than happy to give you permission to proceed. Better still, you can tie your background into your qualifications for the position when the door is open for you to do so.

Now let us take a look at how most people respond to this question. A response for a company officer position might sound something like this:

> Ladies and gentlemen, I have had twelve years with the department. During eight of those years, I was assigned to engine companies. Then I spent four years on truck companies. I have worked in the downtown area, in the high-rise area, in the harbor area, and three years in the busiest area of the city.
>
> During the past two years, I have been acting captain on an engine company and have been in charge of the company at several major emergencies. I have experience on all types of structure fires, fires on the waterfront, and fires in brush areas. In addition to my experience as acting captain, I had three years of experience as an officer in the Navy prior to being employed by the fire department.
>
> My education includes an Associate of Arts degree with 24 units of fire science. I have taken courses in teaching principles and have taught the course Introduction to Fire Science at our local community college. I have also had six courses in computer science.
>
> I am a member of several community organizations, including the Lions Club and the Toastmasters Club. At the present time, I am Vice President of the Toastmaster group. I am married, have two young sons, and am in excellent health. I am qualified and ready to assume the position of company officer.

The presentation is not too bad, and the candidate has a decent background for the position of company officer, but what type of presentation has he made? He has given the board a good summary of his background, and left it up to the board to assume that this background qualifies him for the position. He has not made any attempt to sell his product—himself. Furthermore, he has not done anything to separate himself from his competitors. Most true competitors will have similar backgrounds and experience. Most of them will have had acting-captain experience, most will have worked on various types of companies, most will have responded to various types of fires, most will have taken college courses in fire and computer science, and most will have been involved in some type of community activities. If all candidates make similar presentations, what will the oral board use to separate them? It is important to not just build a pile of bricks (see Figure 6-1).

This brings us to the primary point. In an oral interview, you are a salesperson. You are selling a product, not telling a story. The product you have to sell is yourself. What does a salesperson need to know and do to get his or her customers to buy his or her product rather than the product of a competitor? First of all, the salesperson has to know the needs of the buyer. Second, the salesperson has to know his or her product. And third, one must be related to the other. Of the three, the third one is probably the most important and is, without doubt, the most neglected factor in an oral interview.

An example: Suppose you are selling houses. Before you show your potential buyers any homes, you should chat with them for a while to find out exactly what they want. Suppose they tell you they want a three-bedroom ranch-style house with two bathrooms, a large family room, a formal dining room, and a two-car garage. They do not want to pay more than $200,000. At this point, you have to know the houses available that have all the things they want and which ones you think will most nearly fit their needs. If you try to sell them a $250,000, two-story, extremely modern four-bedroom house, you not only would be foolish, but you would most likely lose your potential customers. When you show them houses that appear to have the items they want, you have to point out other amenities you know will be appealing, such as schools, shopping centers, and so on. But

Figure 6-1 Don't just build a pile of bricks.

you also have to be careful not to oversell or they will feel under pressure and may go somewhere else.

Now let us take a look at your customers' needs. These are not as tangible as a bedroom or a bathroom, but perhaps they can be pinned down. Your customers' needs are the knowledge, ability, experience, and motivation to get the job done in the most efficient manner. Continuing with the company officer's position, seven factors required to get the job done are:

1. The ability to manage a company
2. The ability to supervise people
3. Firefighting knowledge and ability
4. An adequate education
5. The ability to teach

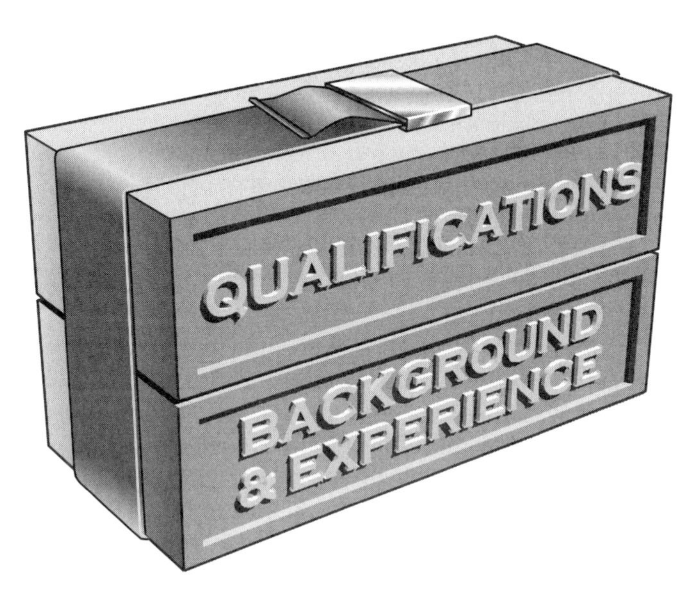

Figure 6-2 Tie your background and experience with your qualifications into a neat package.

6. The ability to work with people
7. The ability to represent the department at public functions

There may be other factors you have in mind, but the preceding list is sufficient to explore the process.

Now let us explore the product. To explore the product means to know yourself. You took a big step toward this in Chapter 3 when you were asked to gather a large amount of material in order to develop a resume. At the end of this chapter, you will be asked to gather more information in order to help you better understand yourself. For now, however, remember that it is absolutely imperative that you know yourself in order to do an effective job of selling yourself in an interview.

Once you know the customer's needs, and truly know yourself, it is necessary to relate one to the other (see Figure 6-2). We can use the presentation of the officer candidate previously given to demonstrate this process.

When determining your customer's needs, you should select four or five qualifications in which you feel you are particularly strong. Keep away from those in which you may have weaknesses. Suppose you are the candidate who made the previous presentation, and you feel you are particularly strong in the areas of supervision and management, firefighting, education, and teaching ability. Your presentation might be changed to one like this.

> Ladies and gentlemen, in my opinion, the four most important qualifications required for a company officer are supervision and management ability, firefighting ability, an adequate education, and the ability to teach. I'd like to take a few minutes to tell you why I am qualified in each of these areas.
>
> First, in terms of supervision and management ability: For the past two years, I have been acting captain of an engine company in the downtown area. During the absence of the captain, I have planned, organized, coordinated, directed, and controlled the operations of

the company. I have maintained the company records and completed all the required reports. I supervised the four firefighters of the company during both routine and emergency operations. This gave me an opportunity to put into effect some of the principles I learned while taking a college course in supervision.

Prior to joining the department, I was a naval officer. In addition to managing the operation of my own command, I indirectly supervised twenty-nine people through my chief petty officers. In that operation, I learned the importance of being fair and impartial while maintaining the discipline required in a military or semi-military operation. My experience as acting captain, together with my naval experience and courses in supervision, have provided me with the knowledge and skills necessary to manage and supervise a fire company.

The second qualification required to be a company officer is to be a good firefighter. This does not mean the ability to lay a line or raise a ladder, but the ability to size up an emergency situation and give the necessary orders to alleviate the problem. I have been in command of the company at a number of fires that could be handled with a single line, but I was also first-in officer at a greater alarm fire in which I turned command of my company over to another member and directed incoming companies to the locations needed. In my evaluation of this fire, I realized that it could not be controlled with the initial response companies and ordered the additional companies I felt were needed. In my request for additional companies, I used an evaluation procedure I had learned in a college course in firefighting.

During my eight years on engine companies, I have been fortunate to work for some excellent fire officers. They taught me the importance of taking time to determine the best place to lay lines and the importance of selecting the proper lines to use. The truck officers I worked for taught me the principles of laddering, overhaul, ventilation, forcible entry, physical rescue, and controlling the utilities. During my twelve years of firefighting, I have learned tactics to use at various types of structure fires, ship fires, refinery fires, high-rise fires, and brush fires. My assignment to the busiest area of the city has provided me with experiences that will benefit me as a company officer in directing the operations of my company.

My educational background includes an Associate of Arts degree with 24 units of fire science and 6 units in computer science. As I mentioned before, one of the courses was in supervision, in which I learned the principles required for managing people. I learned the importance of positive discipline and the need to treat people as individuals. I have been able to put some of these principles into operation in my command of a company.

My courses in computer science have been extremely beneficial in operating the many technical advancements the department has put into operation. A course in firefighting taught me the basic principles of firefighting operations together with the importance of weather, time of day, and type of occupancy. I have had two courses in hazardous materials. On one occasion, I was able to assist my captain at a hazardous chemical spill by sharing with him information I had learned in class about the material involved.

The last qualification I mentioned was the one regarding the importance of being a good teacher. In my opinion, the effectiveness of a company depends upon the quality of the training received by members of the company. I completed a course in teaching principles in which I prepared a course outline and lesson plans. This training taught me the importance of adequate preparation of drills presented by the company officer. I have been able to use the information I learned by teaching a class in fire science at our local college. The knowledge I gained in the course on teaching principles, the application of this information while teaching a course in fire science, and the drills I conducted as an acting captain will greatly assist me in my training program as a company officer.

To sum up, in my opinion, a company officer should be a good manager, a good supervisor, a good firefighter, and a good teacher. He or she should have an adequate education to supplement the knowledge required in each of these disciplines. I am well qualified in each of these areas and ready to assume the duties and responsibilities of a company officer.

Now the difference: In the first presentation, the candidate told a story—a story very similar to that told by other candidates. In the second presentation, the candidate sold himself. He not only told the board the same story as the first candidate, but he related to the board what he had learned from his education and experience which will help him as a company officer. He sold the board on the concept that he was qualified to do the job, while the first candidate left it up to the board to assume he was qualified because he had a given amount of education and experience.

Some people might tell you that the second presentation is too long. This is not so. Time it. You will find it takes about five minutes to present it in a well-paced delivery. This interval is not too long. Also remember that many authorities believe that your final grade may be received within the first five minutes. However, try to confine your presentation to a maximum of about seven minutes. Leave the board some time to ask you questions.

The same type of approach should be used at the entry level and those levels above the company level. The qualifications will be different, but the approach should be the same.

One last word on preparing for the two opening questions: Do not prepare a response and practice is so it will be given in exactly the same words each time. Outline your presentation and know the points you wish to cover and the factors you wish to discuss under each point. Have the outline memorized, but give a little different presentation each time you practice. The board will be quick to recognize a "canned" speech.

Tips for Success Know what you want to say, but give the impression you are thinking about it every step of the way. Learn to make it sound as if your presentation is being given for the first time.

Another factor in your favor of memorizing points to cover and not development of a canned speech: Sometimes the board will interfere with your presentation by asking you another question right in the middle of it. When you are through answering the interposed question, they will tell you to proceed with your presentation. Continuing is much easier if you have the steps rather than the words memorized. For example, you can tell them you have discussed your supervision and management ability and your firefighting qualification and will now continue with your educational background.

AN ORAL INTERVIEW PREPARATION ASSIGNMENT FOR ALL CANDIDATES

Using the information presented in this chapter, prepare yourself a presentation on why you want the position you are seeking. Additionally, prepare a presentation that cleverly ties together your background and experience to the qualifications for the position. In the development of these presentations, use the following outline as a guide.

1. List five categories in which you feel you are the strongest that can be used to demonstrate your qualifications for the position.

2. List four factors from your background that you can use to demonstrate you are qualified in the first category.
3. Write out how you intend to use these four factors to demonstrate that you are qualified in the first category.
4. List four factors from your background that you can use to demonstrate you are qualified in the second category.
5. Write out how you intend to use these four factors to demonstrate that you are qualified in the second category.
6. List four factors from your background that you can use to demonstrate you are qualified in the third category.
7. Write out how you intend to use these four factors to demonstrate that you are qualified in the third category.
8. List four factors from your background that you can use to demonstrate that you are qualified in the fourth category.
9. Write out how you intend to use these four factors to demonstrate that you are qualified in the fourth category.
10. List four factors from your background that you can use to demonstrate that you are qualified in the fifth category.
11. Write out how you intend to use these four factors to demonstrate that you are qualified in the fifth category.

SUMMARY

This chapter provided information on what should be included when explaining to the oral board why you want the position and why you believe you are qualified for the position. It also discussed the importance of tying the two answers into a single package.

If you do your homework well in preparing answers to the questions presented and develop the ability to present them effectively, your oral interview will be successful. Your presentation will probably take place within the most important part of the oral interview—the first five minutes. If you are successful on what is outlined in this chapter, you are further down the path to your goal.

REVIEW QUESTIONS

1. Within how many minutes do most authorities on oral interviews say the grade you receive is determined during an oral interview?
2. What are the two questions of which at least one will probably be asked very early in the interview?
3. List the factors that are considered by people when choosing a lifetime career.

4. If you are applying for a firefighter's position, which five of the preceding factors are most important to you?
5. Where should the fringe benefits and pay be mentioned during your response as to why you want the job?
6. List some of the reasons that employees want to enter the management arena.
7. If you are seeking a company officer's position, which five of the preceding reasons are the most important to you?
8. If the chairperson of the board asks you to tell the board something about yourself, how can you change this so that you will be able to tell the board your qualifications for the position? Prepare your answer as you intend to give it to an oral board.
9. How do most people respond to the question, "What makes you think you are qualified for the position?"
10. What do most people who compete for promotional positions have in common?
11. As a salesperson, what three things do you have to know to get a customer to buy your product?
12. What are some of the factors required to get the job done at the company level?
13. What should be the maximum length of time to use when answering the question "What makes you think you are qualified?"
14. In preparing your presentation to the question of why you want the position and why you are qualified for it, what type of outline should you make?

PROBLEM-SOLVING QUESTIONS

LEARNING OBJECTIVES

The objective of this chapter is to introduce the reader to a process for solving problems. As a result of reading this chapter, the reader should be able to:

- Define a problem.
- Outline the four steps in the problem-solving process.
- Adequately solve problems given to him or her by an oral board.

INTRODUCTION

Many questions asked during an oral interview require that you solve a **problem.** Before you can attempt to solve a problem, you have to know what a problem is. A problem might be defined as the need to make a decision when you have two or more alternatives. If there is only one way to complete a task, then you do not have a problem. Just go ahead and do it.

People encounter a number of problems in their daily lives without ever realizing that a problem exists. Suppose you have eight dollars. It costs that amount to go to a movie you want to see, but there is a special tool you want that costs the same amount. You have a problem. You cannot do both. You make a decision and go to the movie. The alternative you had is generally soon forgotten, and you go about your business. A similar problem may arise again when you have only five dollars and two alternatives on what to buy. However, you will soon solve the problem by making a choice.

The chances are that you give little thought to the mental process you go through when you solve these daily problems. You make a decision and your day goes on as if nothing had happened. However, unconsciously, you went through a problem-solving process. Let's take this unconscious process and use it consciously to help answer questions during an oral interview.

You have a problem. This means you have to make a decision in a situation in which two or more alternatives exist. The mental problem-solving process should proceed as follows.

1. Consider your alternatives. There may be only two but there may be three or more.
2. Consider the probable result of acting on each of the alternatives.
3. Select the alternative with which you can best live once the result manifests itself.
4. Proceed with the alternative you have chosen.

TYPICAL PROBLEM-SOLVING QUESTIONS

Now let us apply this process to several typical questions that might be asked during the interview.

QUESTION: Your company officer tells you to go into a building to make a rescue. When you get inside, you find a two-year old child and a male firefighter. Neither one is breathing. You can only take one out. The one you leave will die. Which one will you take?

SOLUTION: First, it is necessary for you to listen carefully to the question. Candidates continually try to answer this type of question by choosing either the child or the firefighter and say they will take one or the other first and return for the other one. Notice your parameters. You have no second chance. If you take one, the other one dies. The two alternatives then are:

1. Take the child.
2. Take the firefighter.

What will most likely happen if you take the firefighter and let the child die? The press will probably condemn both you and the department for letting the child die. After all, you are being paid to save life and property at the risk of losing your own life. You have sacrificed a small child to save the life of one of your buddies. Not only are you being paid to save the public, but the child had a long life ahead of it while the firefighter had already lived half of his and has enjoyed amenities which the child will never have a chance to experience.

What will probably happen if you take the child and let the firefighter die? You might be written up in the local paper as a hero. While the public may feel sorry for the firefighter who died, the sympathy may not be as great as you think because the public will probably feel that the firefighter was being paid to take this risk. He knew from the day he joined the department that he might lose his life trying to save someone in a fire.

But what will be the reaction from your fellow firefighters? Do you think any of them will ever trust you again at an emergency in which their lives might depend upon your actions, knowing that if you have a choice of saving one of them or a civilian that you will let them die? If any of them ever has to make a similar choice between you and a civilian, knowing the decision you made, do you think there is any chance they would save you?

And what kind of attitude do you expect them to display on a day-to-day basis during normal routine operations? Will all members of the department be pointing their fingers at you and saying, "There's the guy/gal who let one of his/her fellow firefighters die"?

You have a problem. Those are the possible results of taking one alternative or the other. Which one could you most comfortably live with for the rest of your life? How would your thinking have changed if there had been three people in there with two you could not take out and would have to leave to die, and one was a fellow firefighter, one was your mother, and one was a young child?

QUESTION: You are a company commander. The cook tells you that money has been missing from the mess fund. How would you handle this?

SOLUTION: The facts have been presented to you. Someone on your shift or one of the other shifts is taking money from the mess fund. The alternative actions might include:

- Do nothing.
- Explain the problem to all members of your company and seek help in finding the thief.
- Lay a trap for the one taking the money.
- Talk the problem over with the officers on the other shifts and agree on the best action to take.

When considering the alternatives, it should be clear that doing nothing will probably result in the problem continuing. If you explain the problem to all members of your company, you will undoubtedly alert the thief to the fact that his or her actions are known. The stealing may stop for a while but will probably begin again. Laying a trap for the thief may work if the thief is a member of your company but will probably fail if a member of one of the other shifts is involved.

The action to take on this problem is apparent once the alternatives are considered. It is essential that the officers of the other shifts be brought into the solution because the thief could be a member of one of the other shifts. Of course, when you bring in the other officers, the three of you now have a combined problem which will require going through the same problem-solving procedure you faced originally.

The solution here is not greatly different from solutions to other problems. It is surprising how often the answer is simplified once all the alternatives are considered. Incidentally, you will find this problem-solving procedure extremely effective in solving problems in real-life situations as well as in those given to you by an oral board.

Tips for Success When you are in command or faced with an individual problem, if you will take the time to write down the alternatives and consider the probable result of taking action on each of them before you do anything, you will become a more efficient problem solver.

Let's take another problem, one in which the solution seems apparent once the alternatives and the probable results of the alternatives are considered.

QUESTION: You are a rookie firefighter. Your company officer tells you to go on the roof and ventilate. You feel it is not safe to do so. What would you do?

SOLUTION: The three most likely alternatives are to:

1. Obey the order, without question.
2. Refuse to go.
3. Explain to your company officer why you think it is not safe to go on the roof.

You might have seen something the company officer had not. If you obey the order to go, you might injure or kill yourself. If you refuse to go, you will probably be severely disciplined, perhaps fired. The best alternative seems to be to explain to your company officer why you think it is not safe to go. You should be aware that few officers would send a firefighter into a situation where they themselves would not go. If you chose this alternative during the oral interview, you will probably find that another question will immediately follow. It would most probably resemble the following.

Your company officer tells you that he or she has already considered what you saw and for you to get on the roof and ventilate. You now have another problem. You might have been able to avoid the second problem if you had included in your earlier explanation the fact that after explaining to your company officer why you think it is not safe to go on the roof, that you would be guided by whatever he or she says. You know it would then become a matter of his or her judgment against yours, and the officer has a lot more experience than you do.

QUESTION: The fire chief tells you that he or she is considering appointing you to the rank of captain, but if you are appointed, you will have to spend your remaining time on the job on a 40-hour assignment. Would you accept the position under these circumstances?

SOLUTION: Be careful of this type of question. Some people have a tendency to answer it the way they think the oral board wants it answered. With this question, you would probably think the oral board wants you to say you would take the job. Remember, however, that the oral board is not looking for a particular answer. They are analyzing the way you think, evaluating you on the logic of your thought process. Be honest in your response. The two alternatives are:

1. You would accept the promotion.
2. You would not accept the promotion.

Accepting the promotion means you have achieved the goal you have worked so long and hard to obtain. However, you would have to give up so many of the things that firefighters consider worthwhile. You would no longer be working shift duty, with all its fringe benefits. Those extra vacations you used to take would be gone forever. The time you would have with your family would also change. You would miss the challenge of firefighting, the interesting activity that goes on around the engine house, and the camaraderie that exists. But you would be home at Thanksgiving and Christmas. You would have weekends and holidays off. Some people might not object to the change to a 40-hour week. Others would. It is your choice. Make a decision based on the alternative you will best be able to live with, and tell the oral board why you made this choice. Give them every opportunity to understand your thought process.

This explanation just about covers the process that you should consider using for solving problems. There is something important that should be added. How many times have you seen someone do something that you thought was rather stupid? Perhaps later you had a chance to talk to that individual about the incident. The individual explained why he or she did it, and you changed your opinion. Maybe it wasn't stupid after all. Many times when we know why people handle matters in a certain way, their actions seem reasonable, but without knowing the reason, their actions do not make sense. Use this philosophy during the oral interview. After you have made a decision on a problem by considering all the alternatives and the likely result of each, tell the board, "Here is what I would do," and "Here is why I would do it this way."

However, if the board responds with, "But suppose this was the situation?" you may not have covered the problem thoroughly.

DEVELOPING PROBLEM-SOLVING SKILLS

The following section of this chapter is a participation section designed to assist you in developing your problem-solving skills. It is recommended that you complete all of the problems with the reminder that you are developing a thought process, and not particularly solving a problem. When you are asked what your answer would be to the oral board, for maximum benefit, give it exactly as you would to the oral board both in manner and voice and, if possible, give it out loud and tape it. Problems have been provided for both entry-level firefighter candidates and company officer candidates.

Practice Problems for Entry-Level Firefighter Candidates

1. You are a rookie firefighter on probation. The members of the department go out on strike. Will you go with them?
   - **a.** What are your alternatives?
   - **b.** What would be the probable result of each alternative?
   - **c.** Which alternative would you take?
   - **d.** What would be your answer to the oral board?
2. What would you do if your supervisor gave you an order you thought was legally or morally wrong?
   - **a.** What are your alternatives?
   - **b.** What would be the probable result of each alternative?
   - **c.** Which alternative would you take?
   - **d.** What would be your answer to the oral board?

3. Your department has a rule that no firefighter will accept a gift from a civilian at a fire. After extinguishing a fire in a garage, the owner of the garage offers all the firefighters some coffee. Would you accept the invitation?
   a. What are your alternatives?
   b. What would be the probable result of each alternative?
   c. Which alternative would you take?
   d. What would be your answer to the oral board?
4. You have completed your probationary period as a firefighter and now have tenure. The members of the department go out on strike. What would you do?
   a. What are your alternatives?
   b. What would be the probable result of each alternative?
   c. Which alternative would you take?
   d. What would be your answer to the oral board?
5. What action would you take if you place number one on this examination and the chief hires number two instead of you?
   a. What are your alternatives?
   b. What would be the probable result of each alternative?
   c. Which alternative would you take?
   d. What would be your answer to the oral board?
6. Fire department policy is not to enter a burning building without wearing a breathing apparatus. At an extremely smoky structure fire, your company officer checks out the situation and tells you to go in and make a complete search of the building without wearing a breathing apparatus. How would you handle this?
   a. What are your alternatives?
   b. What would be the probable result of each alternative?
   c. Which alternative would you take?
   d. What would be your answer to the oral board?
7. You are taught at the training tower the proper method of raising a ladder. When you are assigned to a company, your company officer tells you to raise the ladder in a manner different from what you were taught. How would you handle this situation?
   a. What are your alternatives?
   b. What would be the probable result of each alternative?
   c. Which alternative would you take?
   d. What would be your answer to the oral board?
8. You receive an alarm of fire and recognize the address as your home. Upon arrival, one of the neighbors tells your company officer that there is still someone in the house. The house is well-involved with fire. Your company officer tells you to take a line and protect the exposures. What would you do?
   a. What are your alternatives?
   b. What would be the probable result of each alternative?
   c. Which alternative would you take?
   d. What would be your answer to the oral board?

9. Your company officer orders you into a building that you know is going to collapse. What would you do?
   a. What are your alternatives?
   b. What would be the probable result of each alternative?
   c. Which alternative would you take?
   d. What would be your answer to the oral board?
10. The chief officer on one of the other shifts is a personal friend of yours. Your company officer calls you in and starts asking you questions about some of the chief's off-duty activities. How would you handle this?
   a. What are your alternatives?
   b. What would be the probable result of each alternative?
   c. Which alternative would you take?
   d. What would be your answer to the oral board?

Practice Problems for Company Officer Candidates

1. You are applying for the position of company officer. The oral board asks you if you would quit our department and take a job as company officer for another department if the pay at the other department had been 50 percent greater. The oral board tells you that the working conditions, pension, fringe benefits, and so on, will remain the same. Would you take the job?
   a. What are your alternatives?
   b. What would be the probable result of each alternative?
   c. Which alternative would you take?
   d. What would be your answer to the oral board?
2. An order comes from headquarters with which you disagree. How would you pass this order on to your crew?
   a. What are your alternatives?
   b. What would be the probable result of each alternative?
   c. Which alternative would you take?
   d. What would be your answer to the oral board?
3. The department policy is that every company will have thirty minutes of exercise every day. You have arrived on your first day at a new assignment. When it comes time for exercises, none of your crew suits up. They tell you they have not been doing "that silly junk." How would you handle this?
   a. What are your alternatives?
   b. What would be the probable result of each alternative?
   c. Which alternative would you take?
   d. What would be your answer to the oral board?
4. You are out on company fire prevention. There is a kindly old woman with a severe hazardous grass problem. She has a very limited income and is unable to do the work herself. How would you handle this situation?
   a. What are your alternatives?
   b. What would be the probable result of each alternative?

   c. Which alternative would you take?
   d. What would be your answer to the oral board?
5. How would you handle a seventeen-year old girl who continues to show up at the fire station, interrupting station activities?
   a. What are your alternatives?
   b. What would be the probable result of each alternative?
   c. Which alternative would you take?
   d. What would be your answer to the oral board?
6. The first female firefighter hired by the department is assigned to your company. There are no separate sleeping facilities or bathroom facilities in the station. How would you handle this situation?
   a. What are your alternatives?
   b. What would be the probable result of each alternative?
   c. Which alternative would you take?
   d. What would be your answer to the oral board?
7. You are driving down the street on your day off and notice that a gasoline tank truck ahead of you is leaking a lot of gasoline. You do not have a cell phone. What would you do?
   a. What are your alternatives?
   b. What would be the probable result of each alternative?
   c. Which alternative would you take?
   d. What would be your answer to the oral board?
8. You are off duty eating dinner in a restaurant and notice that the owner has locked one of the required exits. What would you do?
   a. What are your alternatives?
   b. What would be the probable result of each alternative?
   c. Which alternative would you take?
   d. What would be your answer to the oral board?
9. If the members of the union go out on strike, are you going out with them?
   a. What are your alternatives?
   b. What would be the probable result of each alternative?
   c. Which alternative would you take?
   d. What would be your answer to the oral board?
10. You go into a building to make a search. You find that both one of the members of your crew and your chief officer have collapsed. You can only take one of them to the outside. The other one will die. Which one would you take?
   a. What are your alternatives?
   b. What would be the probable result of each alternative?
   c. Which alternative would you take?
   d. What would be your answer to the oral board?
11. You are out of quarters on fire prevention. You hear a radio report of a dispatch to a fire in another portion of the city. The report includes the fact that people are trapped in the building. The location of the emergency is a considerable distance

from your own district; however, you recognize the address as that of your home. What would you do?

a. What are your alternatives?
b. What would be the probable result of each alternative?
c. Which alternative would you take?
d. What would be your answer to the oral board?

SUMMARY

This chapter was an informational/participation chapter. In it, you were given a system to help you solve problems given to you by the oral board.

You were given a number of problems on which to apply the system suggested. Additional problems on which to practice the concept will be given in other chapters. The system provided will primarily be used in the oral interview in the general questioning and problem-solving portions of the interview.

REVIEW QUESTIONS

1. Define the word "problem."
2. What are the four steps in the problem-solving process?
3. How should you explain your decision to the oral board regarding a problem they gave you?

CHAPTER 8

PERSONNEL PROBLEMS

LEARNING OBJECTIVES

The objective of this chapter is to provide the reader with some food-for-thought regarding personnel problems. After digesting the information in the chapter, and coming to grips with how you would handle certain personnel problems, you should be able to:

- Explain when and if to report the wrongdoing of others.
- Describe how to confront wrongdoers.
- Explain what to do about wrongdoings reported to you.
- Handle theft.
- Limit gambling.
- Deal with those who drink on duty or who are suspected of it.
- Explain how to take disciplinary action.

INTRODUCTION

Personnel problems given during the interview may manifest themselves as problems between you and an individual of the same rank, between you and a subordinate, or between you and a superior. **Personnel problems** are situations in which a conflict exists; that is, the problem as given has two or more alternatives as possible solutions. The thought process for solving the problem is the same as that presented in the preceding chapter. You have to first consider the alternatives. Second, you have to consider the possible result of taking each of the alternatives, and third, you have to choose the alternative with which you will best be able to live.

REPORTING THE WRONGDOING OF OTHERS

One of the alternatives for many of the personnel problems will be that of reporting a person for a wrongdoing. The person is usually one of your fellow firefighters, one of your subordinates, or one of your superiors. You should be aware of some of the reasons people report the wrongdoing of another, because the reason for reporting the incident will probably have some effect on how well you will be able to live with your decision. Although you might be able to think of other reasons, six of the reasons people would report the wrongdoing of others are listed here.

1. *It is the individual's responsibility to do so.* This reason primarily applies to a supervisor who finds it necessary to discipline a subordinate and, in order to do so, has to report the incident through channels.
2. *It is done for revenge.* This reason generally applies to someone who wants to get even with another person. Such reporting is normally done secretly or anonymously.
3. *It is done because the act committed is wrong and the reporting person feels a moral obligation to see that it is corrected.* Many people who report a wrongdoing fall into this category. The act committed goes against the reporting person's principles or someone has done something not permitted and the reporting person does not like it.
4. *It is done for a reward.* The reward may be financial or it may be psychological. Normally, it is the latter. The reporting person feels that he or she will get a "pat on the back" or gain the attention of superiors by reporting the incident.
5. *The wrongdoing might bring discredit to the observing person or to the organization of which he or she is a member if not corrected or eliminated.*
6. *If not reported and allowed to continue, the wrongdoing may do personal harm to the reporting person.*

CONFRONTING THE WRONGDOER

One of the alternatives that might be available to you when you are faced with a situation involving the wrongdoing of another is to confront the individual who committed the act. Following are four methods available for you to do so.

1. You can tell the person you are aware of what he or she did and that you want it corrected. This possibility generally manifests itself in a situation in which the wrongdoer has taken something that does not belong to him or her and can return it without harming anyone.
2. You can inform the individual that you are aware of what he or she did, and you do not want it to happen again. This possibility generally manifests itself in a situation where it is too late to remedy the wrongdoing.

3. You can inform the individual that you are aware of what he or she did, that it is to be corrected, and that if he or she does not correct it, you will report the incident to a superior.
4. You can inform the individual that you are aware of what he or she did, you do not want it to happen again, and if it does, you will report the incident to a superior.

THEFT

At least one of the personnel problems given to you during the interview will probably involve stealing, or alleged theft (see Figure 8-1). The alleged theft is an important consideration. Develop some guidelines for yourself regarding the action you intend to take when theft is involved.

Figure 8-1 You caught this member of your company stealing a bottle of liquor at a fire. What action would you take?

In answering questions regarding this problem, one of your first considerations should be whether or not a theft has actually been committed. Many times the oral board will present you with a situation to which the inference could be drawn that a theft may have been committed but, in actuality, it has not. Be careful of these situations. Before taking any action regarding the alleged offense or before accusing an individual of theft, make sure of your facts.

Also, it is important that you do not confuse theft with poor judgment. Although people may be disciplined for using poor judgment that results in an infraction of a rule, the action taken is generally not as severe as a situation involving theft. One guideline you might consider to help you separate the two is **intent.** In your opinion, did the person take something that belonged to another with the intent to employ it for his or her own personal use, knowing that such action is theft? Or did the person unthinkingly take or use something of another person for his or her own benefit because of poor judgment and in so doing did not think of the act as stealing.

Tips for Success If you are taking the oral interview for the position of company officer, the oral board will be particularly interested in your thought process, whether or not you collect all the facts, and what type of action you will take. Remember, just telling them the action you are going to take when they give you a personnel problem does not tell them what they want to know. They want to know how you think, and what kind of company officer you will probably be.

GAMBLING

Another personnel problem that might be given you will involve **gambling.** By definition, gambling is betting on an uncertain outcome or playing a game for money or property. Most department rules and regulations prohibit gambling on department property or on duty. You as a supervisor are expected to enforce the rules of your department, but you are also expected to use common sense in your relationship with your employees. Every person has to draw the line where he or she will balance common sense with enforcement, and different people will draw different lines at different places. It matters little where you draw your line as long as you use common sense, but it is important that you know where you are going to draw it before preparing answers to questions that will be hurled at you during an interview. For instance, are you going to permit football pools, World Series pools, and other wagers in your command? Are you going to allow your crew to play handball, volleyball, and so on for ice cream for the house? How about that penny-ante card game (see Figure 8-2)? And how about that pinochle game in which there is no money on the table but the players are making marks on a notepad as to who owes whom? If you are going to allow a football pool in quarters, are you going to participate in it? Is your decision logical? Can you justify it both to the oral board and to yourself?

Figure 8-2 Are these members just playing cards for the fun, or are they gambling? As company commander, how would you handle this situation?

DISCIPLINARY ACTION

As you progress through this chapter, you may question whether the objective is to teach supervision principles or to prepare you to answer oral board questions. The objective has been to make you think about what you would do when confronted with certain types of personnel problems. Oral boards love to ask personnel questions. The answers provide the board with a good understanding of the thought process of a candidate. As often stated, this is what they are after, not a particular answer to a particular question.

If you are seeking the position of company officer, many of the situations given to you during the interview will involve you having to make a decision as to whether or not to discipline an employee and, if it is necessary to discipline him or her, what disciplinary action you will use. When solving these problems, keep in mind that the purpose of disciplining is to ensure that the infraction will not occur again. Sometimes a severe tongue-lashing will do the job. At other times, it will be necessary to administer a mild reprimand, and at still other times more severe disciplinary action will have to be taken. Occasionally, it may even be necessary to fire the employee.

When you are preparing for personnel questions that may be asked you during the interview, give considerable thought to what type of action you would recommend if an employee of yours violated any of the rules and regulations of your department. For example, what type of infraction would an employee have to commit before you would recommend that he or she be fired? What would you do about an employee of yours who committed a minor infraction for which you gave him or her a verbal reprimand and a few weeks later the individual repeats the infraction? What is going to be your policy about drinking on the job, an employee reporting for duty with liquor on his or her breath, sexual acts committed on department property, stealing, and the like? What different type of action are you going to take against a good employee as compared to one with whom you have previously had problems? What would be the difference between the action you would take against an employee who stole an item with a value of less than one dollar as compared with one worth ten thousand dollars? Would the penalty you recommend for an infraction of a rule be different if committed by a subordinate or by a superior? The more thought you give to these factors in advance of the interview, the better prepared you will be to answer such questions.

INFORMATIONAL SUMMARY OF PERSONNEL PROBLEMS

Personnel problems can be given to you by the oral interview board whether you are an entry-level firefighter candidate or a candidate for promotion to the rank of company officer. The personnel problems will vary from minor ones to those in which it may be necessary to fire an individual.

Tips for Success Regardless of the magnitude of the problem, it is important that you use the problem-solving method of arriving at a course of action. The more thought you give to these types of problems prior to the interview, the easier it will be to come to a satisfactory solution when answering a question. It is important that you think out the actions you will take in varying situations. Remember that a large proportion of the problems given during the interview involve drinking, theft, and gambling. You should come to grips with each of these issues, and have it established fairly well in your mind what your recommendation for punishment would be in each of these situations. It is also important to keep in mind that the problems will involve subordinates, those of equal rank, and superiors. Think thoroughly about what you really intend to do in various situations, and be prepared to tell the oral board what you would do, and why you would do it in that way.

DEVELOPING PROBLEM-SOLVING SKILLS INVOLVING PERSONNEL PROBLEMS

The objective of this section is the same as that given in the previous chapter. This section also includes problems for entry-level firefighters and company officers. For maximum benefit, when you give your answer by using one of the practicing methods suggested in Chapter 10, give it exactly as you would before an oral board. If possible, give your answer out loud and videotape it. At a minimum, audiotape it.

Practice Problems for Entry-Level Firefighter Candidates

1. About three o'clock in the afternoon, you notice the smell of liquor on your company officer's breath. What would you do?
   a. What are your alternatives?
   b. What would be the probable result of each alternative?
   c. Which alternative would you take?
   d. What would be your answer to the oral board?
2. One of your co-workers has an extremely bad body odor problem. What would you do about this?
   a. What are your alternatives?
   b. What would be the probable result of each alternative?
   c. Which alternative would you take?
   d. What would be your answer to the oral board?
3. You notice that one of the male firefighters you work with climbs out of bed every shift and goes outside. One night you get up and follow him and you find that he is meeting a girl in the parking lot. What would you do about this?
   a. What are your alternatives?
   b. What would be the probable result of each alternative?
   c. Which alternative would you take?
   d. What would be your answer to the oral board?
4. One of the men you work with consistently brags about smoking marijuana off duty. What would you do about this?
   a. What are your alternatives?
   b. What would be the probable result of each alternative?
   c. Which alternative would you take?
   d. What would be your answer to the oral board?
5. One of the women you work with keeps trying to force her religious beliefs on you. How would you handle this?
   a. What arc your alternatives?
   b. What would be the probable result of each alternative?
   c. Which alternative would you take?
   d. What would be your answer to the oral board?
6. After extinguishing a kitchen fire, you notice that one of the firefighters in your company picks up a doughnut and starts eating it. What would you do about this?
   a. What are your alternatives?
   b. What would be the probable result of each alternative?
   c. Which alternative would you take?
   d. What would be your answer to the oral board?
7. At a fire, you see one of your company firefighters pick up some money and put it in his pocket. What would you do about this?

- **a.** What are your alternatives?
- **b.** What would be the probable result of each alternative?
- **c.** Which alternative would you take?
- **d.** What would be your answer to the oral board?

8. Your company officer tells you that your regular relief has called in sick. After being relieved that day, you go to the beach. You see the firefighter who had called in sick playing in the water with his girlfriend. What would you do about this?

- **a.** What are your alternatives?
- **b.** What would be the probable result of each alternative?
- **c.** Which alternative would you take?
- **d.** What would be your answer to the oral board?

9. You are a rookie firefighter. You are in a bar on your day off. You notice that one of the firefighters from your company on the other shift has entered the bar in uniform. He goes up to the bar and orders a drink. What would you do about this?

- **a.** What are your alternatives?
- **b.** What would be the probable result of each alternative?
- **c.** Which alternative would you take?
- **d.** What would be your answer to the oral board?

10. You walk into the locker room one night after dinner and find one of the members of your company taking a drink of liquor from a bottle he has in his locker. What would you do about this?

- **a.** What are your alternatives?
- **b.** What would be the result of each alternative?
- **c.** Which alternative would you take?
- **d.** What would be your answer to the oral board?

11. When parking your car one morning, you see the apparatus operator on your shift back his car into the side of the battalion chief's private car, folding in the fender. The individual then parks his car at the far end of the parking lot. You are the only person to observe the incident. Later, the chief comes into the kitchen where the entire crew is drinking coffee and wants to know who backed into his car. The apparatus operator remains silent. What would you do?

- **a.** What are your alternatives?
- **b.** What would be the probable result of each alternative?
- **c.** Which alternative would you take?
- **d.** What would be your answer to the oral board?

12. You are a rookie firefighter. Upon reporting to work one morning, you notice that someone has opened your locker and used your toothpaste. What would you do about this?

- **a.** What are your alternatives?
- **b.** What would be the probable result of each alternative?
- **c.** Which alternative would you take?
- **d.** What would be your answer to the oral board?

Now that you have had some practice answering personnel problems, it is time to prepare for the practice sessions that will be given later. To do so, write each of the problems that were given you on to a 3×5-inch card and place them in the card file with the other cards you have acquired. Plan to use them in the practice sessions.

Practice Problems for Company Officer Candidates

1. Money has been taken from the food locker. You have investigated the incident and know that one of two members of your crew must be responsible. How would you handle this?
   a. What are your alternatives?
   b. What would be the probable result of each alternative?
   c. Which alternative would you take?
   d. What would be your answer to the oral board?
2. Your immediate chief officer gives you a tongue-lashing for an error caused by an officer on another shift. How would you handle this?
   a. What are your alternatives?
   b. What would be the probable result of each alternative?
   c. Which alternative would you take?
   d. What would be your answer to the oral board?
3. At a fire, you order one of the members of your crew to go on the roof and ventilate. He tells you that in his opinion it is not safe to do so, and he is not going. What would you do?
   a. What are your alternatives?
   b. What would be the probable result of each alternative?
   c. Which alternative would you take?
   d. What would be your answer to the oral board?
4. Your apparatus operator is overweight and getting heavier every day. How would you handle this?
   a. What are your alternatives?
   b. What would be the probable result of each alternative?
   c. Which alternative would you take?
   d. What would be your answer to the oral board?
5. You have a non-cooperative senior firefighter who is consistently getting the rookies to do his work. How would you handle this?
   a. What are your alternatives?
   b. What would be the probable result of each alternative?
   c. Which alternative would you take?
   d. What would be your answer to the oral board?
6. At a fire in a single family dwelling, you notice a member of your company pick up a wallet and put it in his turnout coat pocket. What would you do?
   a. What are your alternatives?
   b. What would be the probable result of each alternative?

c. Which alternative would you take?
d. What would be your answer to the oral board?

7. You are in charge of a working fire in the downtown area. A large crowd has gathered. When the battalion chief arrives, he staggers out of his car and starts shouting orders. It is obvious that he has had a considerable amount to drink. What would you do?
   a. What are your alternatives?
   b. What would be the probable result of each alternative?
   c. Which alternative would you take?
   d. What would be your answer to the oral board?
8. At your new assignment, the long-time permanent cook is also your apparatus operator. He does a fine job of cooking, but you notice that he neglects the apparatus. What would you do?
   a. What are your alternatives?
   b. What would be the probable result of each alternative?
   c. Which alternative would you take?
   d. What would you tell the oral board?
9. You have a senior firefighter who does a good job at fires but performs poorly at drills, primarily because of a lack of preparation. How would you handle this?
   a. What are your alternatives?
   b. What would be the probable result of each alternative?
   c. Which alternative would you take?
   d. What would be your answer to the oral board?
10. You walk out into the parking lot one morning and find one of your firefighters filling a one-gallon can with gasoline from the station pump. What would you do?
    a. What are your alternatives?
    b. What would be the probable result of each alternative?
    c. Which alternative would you take?
    d. What would be your answer to the oral board?
11. You have a firefighter who does outstanding work at drills and is very knowledgeable but performs poorly at emergencies. What would you do?
    a. What are your alternatives?
    b. What would be the probable result of each alternative?
    c. Which alternative would you take?
    d. What would be your answer to the oral board?
12. You walk into the locker room and see a member of your crew taking a drink from a bottle of vodka he had in his locker. What would you do?
    a. What are your alternatives?
    b. What would be the probable result of each alternative?
    c. Which alternative would you take?
    d. What would be your answer to the oral board?
13. You notice that one of your firefighters reports to duty with the smell of liquor on his breath from a previous night's party. What would you do?
    a. What are your alternatives?

  - **b.** What would be the probable result of each alternative?
  - **c.** Which alternative would you take?
  - **d.** What would be your answer to the oral board?

14. You have an African-American firefighter on your company who is occasionally late for work but otherwise does a very effective job. What would you do?
  - **a.** What are your alternatives?
  - **b.** What would be the probable result of each alternative?
  - **c.** Which alternative would you take?
  - **d.** What would be your answer to the oral board?

15. About three o'clock in the afternoon you notice the smell of liquor on the battalion chief's breath. What would you do?
  - **a.** What are your alternatives?
  - **b.** What would be the probable result of each alternative?
  - **c.** Which alternative would you take?
  - **d.** What would be your answer to the oral board?

16. One of the firefighters on your company has an extremely bad body odor problem. The other members of the crew are complaining about him. The man convinces you that he showers frequently and changes his underwear and socks every day. What action would you take?
  - **a.** What are your alternatives?
  - **b.** What would be the probable result of each alternative?
  - **c.** Which alternative would you take?
  - **d.** What would be your answer to the oral board?

17. After a market fire, you find that all the members of your crew have each opened a bottle of soft drink and are also sharing a cake. How would you handle this?
  - **a.** What are your alternatives?
  - **b.** What would be the probable result of each alternative?
  - **c.** Which alternative would you take?
  - **d.** What would be your answer to the oral board?

18. One of the members of your company tells you that your female firefighter smokes marijuana on her days off. What would you do?
  - **a.** What are your alternatives?
  - **b.** What would be the probable result of each alternative?
  - **c.** Which alternative would you take?
  - **d.** What would be your answer to the oral board?

19. Before going off duty, you receive a call from a member of the other shift saying he is sick and will not be able to report for duty. You make the necessary - arrangements to fill the assignment. Later that day, you go to the zoo with your family. You see the man there with his girlfriend. What would you do?
  - **a.** What are your alternatives?
  - **b.** What would be the probable result of each alternative?
  - **c.** Which alternative would you take?
  - **d.** What would be your answer to the oral board?

20. You take your monthly reports to the battalion chief one night after dinner. Upon entering his office, you notice a bottle of whiskey together with a glass half full of whiskey and water on his desk. What would you do?
    a. What are your alternatives?
    b. What would be the probable result of each alternative?
    c. Which alternative would you take?
    d. What would be your answer to the oral board?
21. You go out to the parking lot one night at about ten o'clock and notice that one of your male firefighters has a woman in the back seat of his car. She is completely undressed. What would you do?
    a. What are your alternatives?
    b. What would be the probable result of each alternative?
    c. Which alternative would you take?
    d. What would be your answer to the oral board?
22. One of the men from the other shift comes to you and tells you that a member of your company has been dating his wife, and he is furious. What would you do about this?
    a. What are your alternatives?
    b. What would be the probable result of each alternative?
    c. Which alternative would you take?
    d. What would be your answer to the oral board?
23. One of the members of your company is going through a divorce. This is having an adverse affect on both her routine and emergency performances. What action would you take?
    a. What are your alternatives?
    b. What would be the probable result of each alternative?
    c. Which alternative would you take?
    d. What would be your answer to the oral board?

Now that you have had some practice answering personnel problems, it is time to prepare for the practice sessions that will be given later. To do so, write each of the problems that were given to you on to a 3×5-inch card and place them in the card file with the other cards you have gathered. Plan to use them in the practice sessions.

SUMMARY

This chapter was an information/participation chapter. In it, you were given different types of situations that could develop into personnel problems. You were also given a number of personnel problems to solve. Each of those given to you might be asked of you during the oral interview. Interviewers have a habit of asking personnel questions as this type of question provides the board with an understanding of how you think. Developing a procedure for solving these problems and thinking out ahead of time as to how you would answer each question will greatly assist you in preparing for your interview. You will be asked to solve additional problems of this type during your practice as outlined in chapter 10.

REVIEW QUESTIONS

General Questions

1. Between what ranks of people may the personnel problems given during the oral interview manifest themselves?
2. What thought process should be used to solve personnel problems?
3. What are some of the reasons that people report the wrongdoing of others?
4. What are some of the methods available for confronting a person who has committed a wrongdoing?

Company Officer Candidate Questions

1. How would you define stealing?
2. What penalty would you recommend if a member of your crew stole:
   - **a.** a bottle of soft drink?
   - **b.** a bottle of whiskey?
   - **c.** a $10,000 ring?
3. If you did not recommend the same penalty for each of the infractions given in question 2, why did you differentiate?
4. How would you define gambling?
5. What penalty would you recommend if you caught a member of your crew gambling?
6. Does a football pool fall within your definition of gambling? Why or why not?
7. Would you allow the members of your company to play handball for ice cream for the house? Why or why not?
8. Would you allow members of your company to play cards if they were making marks on a paper showing who owed whom but there was no money on the table? Why or why not?
9. What do you intend to do if a member of your company comes to work with liquor on his or her breath?
10. Would you act in the same way if one of the other company officers in the engine house or the battalion chief comes to work with liquor on his or her breath?
11. What do you intend to do if you find a member of your crew drinking liquor on duty?
12. Would you do the same thing if you observed the battalion chief drinking on duty? Why or why not?
13. What types of disciplinary action do you intend to take other than an oral reprimand and firing an employee?
14. What do you intend to do if a wrongdoing that involves something one of your crew did off-duty is reported to you?
15. What do you intend to do if a wrongdoing that involves something a member of your crew has done on-duty is reported to you?

CHAPTER 9

FIREFIGHTING QUESTIONS

LEARNING OBJECTIVES

The objective of this chapter is to introduce company officer candidates to principles that they can use to answer firefighting questions from the oral board. This information is specifically related to major fires in which the candidate is placed in command of the emergency. As a result of completing this chapter, a company officer candidate should be able to:

- Explain the actions required when taking command.
- Describe what to include in the initial report to the dispatch office.
- List the major functions to be performed at a fire.
- Explain to which company the major functions are assigned.
- Discuss how to establish a system for estimating the personnel and equipment needed at a fire.
- Describe a method for directing incoming companies.
- Explain what to do when placed in command of an unusual situation.

INTRODUCTION

This chapter is different from the others. First, it has been written solely for those preparing for the position of company officer. The reason for this is that entry-level firefighter candidates are not expected to know much about firefighting itself. If they are asked anything about a fire at the oral interview, it will most likely involve a personnel problem.

But that is not really what makes this chapter different. The difference comes in the jump that is made from previous questions for company officer candidates regarding situations they might encounter in their day-to-day operation of a company. This is not so in the area of firefighting. Here, a candidate will be asked a question about a situation he or she may run into only once or twice during his or her career as a company officer. At no point during your interview will you be more thoroughly tested on the preparation you have made for advancement than you will during the questioning on firefighting. The goal of this interview is to take a candidate with relatively little experience and expect him or her to know all the answers regarding firefighting operations. Such candidates might have had only five or six years in the department, may have held a 2 1/2-inch line in his or her hand at a fire only once or twice if at all, might have controlled a small line at fires only a dozen or so times, might never have vented a building at a greater alarm fire, and might never have been in charge of a fire company at a single incident. Yet, during the interview, the board will put the candidate in charge of a serious, though hypothetical, fire or emergency, just to see how he or she would perform.

The reason for this practice is that the **standard operational procedures (SOP)** of most fire departments direct that one of the first company officers to arrive at a fire is in command until relieved by a superior officer. Therefore, the oral board will place this inexperienced potential officer with a company at a major fire in which the safety of the whole situation depends upon his or her decisions, tell him or her that the first-in chief's car has broken down en route to the fire and, consequently, a chief officer will not arrive on the scene for a long time, and asks this potential officer how he or she would handle the situation.

> **Tips for Success** A company officer candidate should be prepared to handle challenging situations, regardless of the frequency or the likelihood that the event will occur.

Undoubtedly, the preceding warning painted a pretty dismal picture. However, things are not as bad as they seem. Remember that all the officers in your department at one time or another went through the same process you will have to endure, and they seem to have survived. Preparing to take command of a fire is really a matter of learning firefighting tactics and applying them on the fireground. It is assumed that if you have gotten far enough to be taking an oral examination for a company officer's position, you have taken a course in firefighting tactics or at least done enough self-studying to be fairly familiar with the principles involved. You should be familiar with the **National Incident Management System (NIMS),** the **Incident Management System (IMS),** the term **RECEO VS** (an acronym for the seven basic firefighting strategies: rescue, exposures, confinement, extinguishment, overhaul, ventilation, and salvage), and the term **Rapid Intervention Team (RIT).** You should also be aware of the importance of reviewing each of these prior to reporting for an oral interview. The more experience you obtain in commanding units on the fireground, the more confident you will become. However, the basic tactics to be applied do not change. Tactics are learned in classroom situations, by self-study, and on the fireground. Although the basic procedures for laying lines, raising ladders, overhauling, and so on might vary slightly from department to department, the overall objectives vary little from one area of the country to the other. Placing you in the position of Incident Commander seems very different from riding the jump seat. However, in some respects,

Figure 9-1 An Incident Commander needs an adequate number of staff assistants at a large fire.
Courtesy of Rick McClure, Los Angeles Fire Department.

it makes your job easier when you are answering questions from the oral board. If you were given fire problems to solve at the company level, it would require that you think in small details such as what size of line to lay, how much hose to lay out, what ladder to select to reach a certain window, and the like. At the Incident Commander level, you think in terms of functions that have to be performed, what types of companies are required to perform the functions, and then of ordering company officers to get the job done. You have to consider such factors as the number of RITs that will be required, the amount of staff assistants you will need, and where all these people should be placed (see Figure 9-1). It is necessary that you be able to adequately judge the functions that have to be performed, where they have to be performed, when they have to be performed, and how many personnel and how much equipment it will take to get them done.

The following information is not being given to you for the purpose of teaching you firefighting tactics or strategy. You already know them. The objective is to review some of the factors that you should consider in organizing your thoughts for answering questions on taking command at large fires.

TAKE COMMAND

Oral boards will normally give you a firefighting situation in which you are the Incident Commander at a major or unusual emergency (see Figure 9-2). As Incident Commander, you are responsible for taking command of the emergency. When you take command, you

Figure 9-2 Regardless of the size or complexity of the fire, TAKE COMMAND.
Courtesy of Rick McClure, Los Angeles Fire Department.

assume the responsibility for establishing command, directing incoming companies, assessing personnel and equipment needs, and ordering additional units if required.

At a structure fire, normally the best place to set up the command post is outside of the collapse zone directly in front of the building. To direct incoming units and maintain contact with the dispatch office, it is essential that you have a radio at your command. It is also important to remember that you cannot direct incoming companies if you are on the end of a hose line inside the building. Consequently, it is good practice to turn command of your company over to your acting officer and give him or her directions as to what you want done. Keep to your apparatus so that you have a place from which to operate and a means of remaining in contact with the dispatch office and the incoming and working companies.

Figure 9-3 How many companies do you estimate you will need on this fire?
Courtesy of Rick McClure, Los Angeles Fire Department.

INITIAL REPORT

As soon as possible after arriving on the scene, you should size up the situation (see Figure 9-3) and report to the dispatch center as to what you have observed. The report to the dispatch office is also important to responding companies. Your initial evaluation should concentrate on the life problem, where the fire is, where it is going, and what is needed to stop it. At a minimum, your initial report should include:

- where you are
- what you have
- what you need

The standard operating procedure for various departments will differ in regard to how and what should be included in reports to the dispatch center, but a report might go something like this:

> Dispatch from Engine 10. At 1012 North Main, we have several well-involved buildings. The fire itself is in the defensive mode, but there are several serious exposures to the west. Give me five more engine companies and two more trucks. What companies am I receiving on the initial dispatch?

It is assumed that you know the number of companies and the staffing you will receive on the initial alarm. If you do not, you should ask the dispatch center for this information prior to requesting additional companies. Otherwise, you will not know how many additional companies to request. The reason for asking what companies you are receiving on the initial alarm is to provide you with information so that you can start placing the responding companies where they are needed.

FIREGROUND OBJECTIVES TO BE PERFORMED

Sometimes we become so involved in a process that we do not take the time to sit down and thoroughly think out the factors that make up the process. There is no doubt that you have at one time or another been exposed to all the objectives that have to be performed by the fire department on the fireground. You are also aware that the first priority at any fire is search and rescue. However, it may not hurt to review these objectives and put them into an order that can help you in your preparation to take command of a fire. While there may be other objectives to be performed on the fireground, the following is a list of the major ones with which you will probably become involved in order to organize your thoughts regarding answering oral board questions. This list is a little more extensive than RECEO VS; however, if you have been trained to use RECEO VS and are satisfied with it, use it instead of the following:

- search and rescue
- fire extinguishment
- protecting exposures
- laddering, both the fire structure and the exposures
- overhaul
- ventilation
- forcible entry
- controlling the utilities
- emergency medical care

In outlining the objectives that need to be performed at a fire, it is recognized that the majority of fire departments do not have the personnel or the resources that larger departments do. The discussion on what is needed is based upon the ideal and not necessarily what a candidate's department is able to supply. It is therefore necessary for you to base your answer to the oral board on what your department has or is able to receive from mutual aid departments and not on the ideal. While the ideal is being discussed in this chapter, you should base your actions on the reality of the situation.

A quick review of this list of objectives should indicate to you that with the exception of overhaul, it is quite possible that all of the objectives may have to be performed simultaneously in the early stages of the fire. If any of the objectives need to be performed but are not, the total loss from the fire will most likely increase. Although physical rescue always takes first priority at an emergency, all of the other objectives are important for success in fire control. Consequently, the Incident Commander of the fire (who will be you) must give early thought to these and make sure that sufficient personnel and equipment are on the scene, or on the way to the scene, to achieve the objectives required adequately.

Also, it is important to consider auxiliary factors such as the need for a RIT and required operational standards such as the two/in, two/out concept.

To ensure that all the functions that have to be performed are completed, fire departments are organized in such a manner that the responsibility for carrying out each of these functions is assigned to a particular firefighting unit. Different departments use different methods for assigning functions based upon their resources, but most fire departments that have the resources use

- engine companies:
  - extinguish the fire
  - protect the exposures
- truck (or ladder) companies:
  - search and rescue
  - laddering
  - overhaul
  - ventilation
  - forcible entry
  - controlling the utilities
- rescue (or squad) companies:
  - emergency medical care
  - search and rescue

At one time, many large departments had separate salvage companies staffed for the purpose of providing the required salvage operations. Very few now do so. However, salvage operations at a fire are extremely important and must be given early thought by the Incident Commander. If the truck company operations at a fire are limited, this company can be assigned the responsibility for salvage operations. Otherwise, as the person in charge of the fire, you should consider requesting additional engine companies that can be used for performing this operation.

Now let us take a look at some methods you can use to help determine the number of companies and personnel you will need to control the fire.

ESTIMATING THE NUMBER OF COMPANIES NEEDED

Most officer candidates who are given firefighting situations at oral interviews make the same mistake. The fire situation given to them almost always requires more companies and personnel that was assigned on the initial response. The answer generally given to the oral board by a candidate is that he or she would ask for a second alarm assignment, or perhaps a third alarm assignment. If you do this, in most cases, you are really saying that you realize that additional help is needed, but you are not really sure of the amount. As a result, you are going to get more help and let the chief decide the total amount really required when he or she arrives on the scene. This action is a mistake. If you are the Incident Commander, take command. The general principle you should follow is to ask for the number of companies you think you will need, and one additional engine company. Remember that there is always an excuse for ordering too many companies, but none for under-ordering.

Tips for Success If you are going to fail, fail safe.

Always consider the fact that if the fire is of any size (and it will be), you will generally need an engine company downwind to check for flying brands. You will most likely also need an engine or truck company for the RIT and perhaps another to perform salvage operations.

The problem with most candidates is that they have never given any thought regarding a system for determining the number and type of companies needed. This is understandable, as it is not a very important issue when you are riding the jump seat. However, when you are planning for an oral interview for the company officer's position, you have to think ahead to when you will be in charge of the fire.

Although many officers work successfully by using their intuition and experience on the fireground,

Tips for Success Most good officers use some method of estimating the number of companies needed at the emergency.

It would be difficult to label any one method as the best, as there are several ways to fight a fire successfully. The important point is to have a system that you can use to practical advantage, and one you can use to explain to the oral board how you arrived at the number of companies you estimated as being required to handle the emergency. Do not forget that the oral board will be evaluating your thought process and your organizational ability, and will not quibble with you as to whether you should have ordered ten more companies rather than eight.

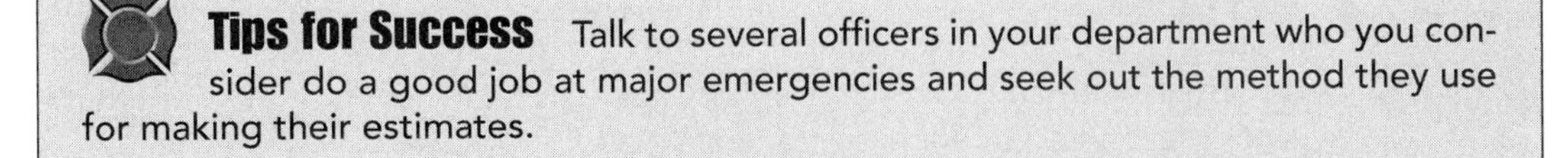

Tips for Success Talk to several officers in your department who you consider do a good job at major emergencies and seek out the method they use for making their estimates.

Don't be surprised to find out that none of them has a method, particularly if you are in a department that seldom has a major emergency. However, if they do have guidelines they use, you might find that by talking to them and then considering the guidelines offered here, you may be able to develop a system you will feel comfortable using and one you think will work well for you.

Following are some guidelines for your consideration. You may or may not be able to use them depending a great deal on the manner in which your department operates and is staffed. Consequently, consider them only as guidelines and not as hard and fast standards.

Probably the best method of determining the amount of fire flow you will need for extinguishing and controlling is to use the formula developed by the National Fire Academy. For a single floor, the NFA formula is to determine the square footage of fire involvement and divide it by 3. For example, if the involved area is 30 feet by 50 feet, the square footage would be $30 \times 50 = 1500$. Divide the square footage by 3, and your fire flow would be 500 gallons per minute. There are additional mathematical calculations that can be used for determining the fire flow required to protect the exposures and for situations in which more than one floor is involved. It is strongly recommended that you become familiar with all the calculations and be capable of mentally determining the total fire flow you will need. Knowing the required fire flow is essential for determining the number of engine companies that will be required on the fire.

In general, a properly staffed engine company is capable of achieving one of the following:

1. Handling a 2-inch or $2\frac{1}{2}$-inch line
2. Staffing a 2-inch or $2\frac{1}{2}$-inch line reduced to two smaller lines
3. Placing a heavy-stream appliance into operations

In certain types of fires, such as those in high-rise buildings, basements, and so forth, that will extend over a considerable period of time, it will be necessary to relieve the firefighters on the line because of the limited time available while using breathing apparatuses. This is important not only from a firefighting standpoint, but more so as a safety factor for the firefighters. These situations generally require three engine companies to staff a working line: one on the end of the line, one waiting at the entrance to relieve the firefighters on the end of the line, and a third replacing breathing apparatus bottles.

At structural fires, a truck company is needed for approximately every three engine companies in order to carry out the responsibilities assigned to the company. If the life hazard is extraordinary, two truck companies to every three engine companies may be needed. If a department does not have any truck companies, an engine company should be dispatched to assume the duties of the truck company.

At least one company doing salvage work should be assigned whenever the fire is extensive and is located above the first floor.

Emergency medical care capabilities should be available at all working fires. These capabilities should be in the form of companies assigned this responsibility only. Rescue companies may be used and ambulances should be ordered if any possible need exists.

Boat companies are seagoing engine companies. They may be able to handle more lines, depending on their size and staffing.

Crash or rescue companies located on airports are similar to engine companies, but generally use light water or Class A foam instead of water to achieve their objectives. They also have the primary responsibility for the life problem during the early stages of the fire.

Knowing the capability of an engine company and having guidelines for determining the number of truck companies, salvage companies, and rescue companies required provides a foundation for the development of a system for determining the companies needed at the emergency. One of your first thoughts should be to determine quickly where you want lines or heavy-stream appliances for the dual purposes of protecting exposures and for fire extinguishment. The number of engine companies needed will equal the sum of the lines required and the heavy streams needed, plus one additional. Use your ratios for determining the number of truck companies, companies performing salvage work, and rescue companies required. If any special equipment is needed, add it to your list. Then request what you want.

DIRECTING INCOMING COMPANIES

There may be another question immediately after you tell the oral board how many companies you need to control the fire: "What are you going to do with all those companies?"

Directing incoming companies requires that you consider the objectives that need to be achieved and assign companies on a priority basis. Saving life takes first priority. However, many times the best method of keeping the loss of life to a minimum is to make a quick, direct attack on the fire.

Sometimes department policy will affect your decision as to where to place the first-alarm companies. For example, many departments have a policy that the first or second engine company should lay into the sprinkler inlet of a sprinkler-equipped building, or the first- or second-in engine company should lay into the exterior standpipe of a multiple-story building that has a standpipe system. Know your department's firefighting procedures, and follow them in your initial placement of companies. Remember, however, that common sense should prevail whenever there is a conflict.

A basic principle of firefighting is to locate, confine, and extinguish. With the pressure you are going to be under during your first command of a major fire, this principle should be kept in mind when considering the placement of companies. Most inexperienced officers will remember to protect exterior exposures, but many fail to consider the need to get lines rapidly above the fire in multiple-story buildings in order to stop the upward spread of the fire. Do not forget to consider the upward spread of the fire early when you are establishing priorities.

Other than evacuation of those people in the building or adjacent buildings who are in danger, priority of assignment generally refers to the placement of engine companies for the purpose of confining and extinguishing the fire. You should make sure you have a sufficient number of engine companies on the scene in order to carry out these priorities. Consider where the fire is likely to be by the time the requested companies arrive on the scene, and not where it is when you make your request.

Once you have determined the number of engine companies you want for fire extinguishment, the number you want to protect the exposures, and a priority for placement, you can start directing companies. Keep your orders simple.

- "Engine three, get a line above the fire."
- "Engine five, protect the exposure on the north side."
- "Truck three, evacuate the third and fourth floors."
- "Engine two, take the back of the building."

REQUESTING ADDITIONAL COMPANIES

The thought that goes through the minds of many company officer candidates who are given large fires to command during the oral interview is that their department does not have the capability of providing either the personnel or the equipment needed. These candidates are probably correct. However, you are in command of the situation and it is up to you to request what is needed.

There are not many departments in the United States that have the capability of controlling major emergencies on their own, and still provide adequate coverage for the rest of their city. Because mutual aid is available in most parts of the country, when you are given a problem during the interview, it is not your responsibility to determine where the companies are coming from. Your responsibility is to determine what is needed and to request it. Do not hesitate to tell the oral board that you will do the best you can with what you have if they inform you that the companies you requested are not available.

UNUSUAL SITUATIONS

Occasionally, the oral board will put you in command of a situation with which you are totally unfamiliar. The situation may involve chemicals or other unfamiliar materials, or it may be that you will be given a problem in an area that requires special knowledge you do not have. There are no hard-and-fast rules for handling these problems, but there are some guidelines you might consider.

The general rule is not to do anything that might increase the problem or that would place civilians or your personnel in jeopardy unless there is an immediate life problem. In most cases, these situations require that you obtain more information before doing anything. For example, if you are faced with an emergency involving chemicals with which you are unfamiliar, keep everyone at a safe distance until you obtain the information you need. The dispatch office may be able to get this for you or at least may be able to get hold of someone who knows the answer. Of course, if your department or a mutual aid department has a hazardous materials squad, then request the unit and wait for it before taking further action.

At other times, you will be placed in command of a situation in a district that has special problems, and you do not have the experience or knowledge to handle the situation effectively. For example, you may move up to a waterfront as a result of a mutual aid request. While there, you may be presented with a fire in the engine room of a ship in which the best method of attack is through the shaft alley. It may be that you have never heard of a shaft alley. If you order a firefighting company into the engine room through the regular entrance, you will place all the firefighters in jeopardy. In these situations, do not hesitate to take an experienced officer from the district under your wing and be guided by his or her advice. There is no disgrace in asking for help when you do not know the answer. In fact, common sense tells you it is the only thing to do.

Although it more nearly approaches a typical situation rather than an unusual one, you should be careful about grossly overestimating the potential of a problem. Knowing that you expect them to give you a very serious problem, the board will sometimes give you a situation that should be handled with the first-alarm assignment just to see if you will overreact. If the problem given you is of such a nature, take command, but handle it with the companies you have, if you believe that those are all you need.

DEVELOPING A THOUGHT PROCESS FOR ANSWERING FIREFIGHTING QUESTIONS

Here are some problems that have been designed to help you think your way through firefighting or other emergency situations when you are placed in command. Answers are not provided because there are no right or wrong answers. If there were, it would imply that there is only one way to handle the situation. This is not true. The problems, therefore, have been designed to help you develop a thought process, and to help you organize your thoughts into a logical method for getting the job done. If you are serious about improving your oral interview score, it is important that you complete these exercises. However, if the problem given is one that could not possibly occur in your city or community, do not waste your time on it.

PRACTICE PROBLEMS

On all of the problems given to you in the remainder of this chapter, assume that your arrival at the fire places you in the position of Incident Commander in accordance with the standard operating procedures of your department.

PROBLEM: As a newly appointed company officer, you have been assigned to a truck company in the primary industrial area of your city. The area consists of closely built structures with little open space. At three o'clock on a Sunday morning you are dispatched to a reported structure fire. When you arrive on the scene, you find a 100×200-square-foot, three-story brick manufacturing plant with a major portion of the second floor well-involved. Fire is extending out most of the windows on the second floor. Describe to us what you would do.

In addition to writing out your response to the oral board:

1. Write out your initial report to the dispatch office exactly as you would give it.
2. After turning your company over to your acting officer, what would you assign the company to do?
3. Where would you set up your command post?
4. List your first five priorities.
5. How many engine companies do you think you will need for fire extinguishment?
6. How many engine companies do you think you will need to protect the exposures?
7. How many truck companies do you think you will need?
8. How many companies doing salvage work do you think you will need?
9. How many rescue companies do you think you will need?
10. Do you think you will need any special equipment or companies? If so, what?
11. Where will you place the first engine company to arrive?
12. Where will you place the second engine company to arrive?
13. Explain where you intend to use the other engine companies you requested.
14. What duties would you assign to the first truck company to arrive?
15. What duties would you assign to the second truck company to arrive?
16. Take command and write out your directions to the first five companies exactly as you would give them.

PROBLEM: As an engine company officer, you arrive at a thirty-story office building in the downtown area. There is a considerable amount of smoke and fire coming out of the windows on the twenty-fifth floor. Tell us how you plan to handle this situation.

In addition to writing out your response to the oral board's question:

1. Write out your initial report to the dispatch office exactly as you would give it.
2. After turning your company over to your acting officer, what would you assign the company to do?
3. Where would you set up your command post?
4. List your first five priorities.
5. How many engine companies do you think you will need for fire extinguishment?
6. How many engine companies do you think you will need to protect the exposures?
7. How many truck companies do you think you will need?
8. How many companies doing salvage work do you think you will need?
9. How many rescue companies do you think you will need?
10. Do you think you will need any special equipment or companies? If so, what?
11. Where would you place the first engine company to arrive?
12. Where would you place the second engine company to arrive?

13. Explain where you intend to use the other engine companies you requested.
14. What duties would you assign the first truck company to arrive?
15. What duties would you assign the second truck company to arrive?
16. Take command and write out your directions to the first five companies to arrive exactly as you would give them.

PROBLEM: At four o'clock on a cold Tuesday morning, you arrive at a five-story apartment house. There is a considerable amount of smoke coming out of the windows on the third and fourth floors. How are you going to handle this?

After writing out the response to the oral board's question:

1. Write out your initial report to the dispatch office exactly as you would give it.
2. Estimate the number and types of companies you would need and request to handle this situation.
3. Explain where you would use each of these companies and for what purpose.
4. What is the total number of personnel you feel will be required for both fire suppression and staff support?
5. Which arriving company would you use for the RIT?
6. After establishing the saving of life as your first priority, what would be your second?

PROBLEM: As a company officer, you and your crew are moved on a mutual aid response into an engine house that is located a half-mile from a large refinery. Five minutes after moving into the engine house, you are dispatched to a reported fire in the refinery. As you roll out the door and turn toward the refinery, you can see that there are two and possibly three 80,000 barrel tanks involved with fire. You will be the Incident Commander when you arrive. How are you going to handle this situation?

After writing out your response to the oral board's question:

1. Write out your initial report to the dispatch office exactly as you would give it.
2. How many engine companies do you estimate will be required to handle this emergency?
3. What will be your first priority at this fire?
4. How long do you estimate it will take to extinguish this fire?
5. Will you need any special equipment for this situation? If so, what type and where would you plan to use them?
6. Are there any unusual needs that you estimate you will have to plan for at this situation?

PROBLEM: As an engine company officer, you respond to a freeway incident at four o'clock on a Friday afternoon. Upon arrival, you find a tank truck containing liquid hydrogen overturned. There is a hole in the side of the tank that is allowing the escape of a considerable amount of the material. As Incident Commander, how are you going to handle this situation?

After writing out your response to the oral board's question:

1. Explain what you consider is the biggest problem you are facing.
2. How are you going to handle this problem?
3. How many companies and of what type do you believe you will need to handle the overall situation?
4. If you decide to use water, where are you going to get it?

The following fires use photographs in lieu of an oral board member's description of the problem facing you. While this technique has not been used too often by oral boards, with the increased use of technical advances by the fire service, it would not be surprising to see an increase in its use. Regardless, the practice you receive in answering the questions will in no way harm you in your preparation for handling firefighting problems before an oral board.

Figure 9-4 *Courtesy of Rick McClure, Los Angeles Fire Department.*

PROBLEM: You are the Incident Commander at the fire shown in Figure 9-4. The photo shows what you see when you arrive.

1. Give your initial report to the dispatch office.
2. How many engine and truck companies do you feel will be required to control and extinguish the fire?
3. Where would you put them?

Figure 9-5 *Courtesy of Rick McClure, Los Angeles Fire Department.*

PROBLEM: You were the first officer to arrive and are the Incident Commander at the fire shown in Figure 9-5. The photo shows the fire as it looked when you arrived.

1. What would be your initial report to the dispatch office?
2. How many engine companies would you request?
3. Where are you going to place your company and the first engine company to arrive?
4. How many truck companies would you request?
5. How many other types of companies would you request?
6. What instructions would you give to each of the arriving companies?

Figure 9-6 *Courtesy of Rick McClure, Los Angeles Fire Department.*

PROBLEM: You were the first officer to arrive and the Incident Commander at the well-involved attached garage fire shown in Figure 9-6.

1. What would be your initial report to the dispatch office?
2. How many engine companies would you request?
3. Where would you place your company and the first engine company to arrive?
4. How many truck companies would you request?
5. How many other types of companies would you request?
6. What instructions would you give to each of the arriving companies?

Figure 9-7 *Courtesy of Rick McClure, Los Angeles Fire Department.*

PROBLEM: You became the Incident Commander when you arrived at the fire shown in Figure 9-7. The photo shows what the fire looked like when you arrived.

1. What would be your initial report to the dispatch office?
2. How many engine companies would you request?
3. Where would you place your company and the first engine company to arrive?
4. How many truck companies would you request?
5. How many other types of companies would you request?
6. What instructions would you give to each of the arriving companies?

Figure 9-8 *Courtesy of Rick McClure, Los Angeles Fire Department.*

PROBLEM: As a result of a mutual aid move-up, you are the first to arrive at the wildland fire shown in Figure 9-8. Your arrival places you in command of fire operations. You have never been on a wildfire, and this one is located above a residential area. How are you going to handle this situation?

SUMMARY

This chapter was devoted 100 percent to handling firefighting problems. A review of firefighting principles was presented and a number of fire problems were offered to help you develop a thought process on how to handle fire problems when you are the Incident Commander.

There is a lot to think about when you are the Incident Commander in command of a fire. However, the preceding information should provide you with knowledge as to how to handle the problems. The best way to learn is to practice on some typical problems that might be given to you during the oral. Keep in mind that you will generally be graded on your thought process and your organizational ability, not on exactly how you handled the fire. If you can demonstrate to the oral board that you have thought through the problem, have established a priority system for getting the job done, and have ordered sufficient personnel and equipment to do the job, you cannot go too far wrong.

PRACTICE, PRACTICE, AND PRACTICE

LEARNING OBJECTIVES

The objective of this chapter is to introduce the reader to the tools and methods that are recommended for completing the practice phase of the preparation for the oral interview. Upon completing this chapter, the reader should be able to:

- Name the tools that are required for the practice sessions.
- Explain the various methods that should be used for practicing.
- Complete a stack of 3×5-inch cards that can be used in the practice sessions.

INTRODUCTION

Congratulations! You have reached the final step in your preparation to appear before an oral board. Now it is time to

Tips for Success Practice, practice, and practice.

You have been exposed to potential opening questions, ways in which to handle problems, how to handle personnel problems, and if you are an officer candidate, firefighting problems.

You have gathered all the information you think might be needed during the interview. A short presentation as to why you want the job has been prepared. This presentation has listed some of the major points you wish to discuss and not word-for-word what you will say. You have also used the same process in preparation to talk about why you are qualified for the position, and you have carefully tied this together with your background and experience. You have prepared a similar presentation on what duties are performed by an individual assigned to the position you are seeking. The oral procedure has been reviewed and the things that should and should not be done during the interview have been learned. You have also learned the procedures for answering questions. As previously mentioned, the time has arrived to practice, practice, and practice.

TOOLS REQUIRED FOR PRACTICING

Before starting to practice, you should gather all the tools required for obtaining the maximum effectiveness out of the practice sessions. Following is a list of four tools that are recommended.

1. *A stack of 3×5-inch cards with a question written on each one.* Before you practice, take all the questions you think might be asked and write them on 3×5-inch cards. If you carefully followed the instructions in this book, then you should already have a large stack of these cards. Having each question on an individual card will give you the ability to shuffle the cards so that the questions are presented in a different sequence each time you use them. Then write the questions at the end of this chapter on cards to add to the stack. The completed stack will be used by you and others to ask questions for you to answer. This stack of cards will be the most important tool you need for practicing.
2. *A camcorder.* A camcorder will be the second most important tool you will need for practicing. The camcorder will be invaluable for you to observe your performance and the manner that you answer questions.
3. *Some type of voice recorder.* The voice recorder will be used to record your presentation whenever the situation for practicing is one in which a camcorder cannot be used.
4. *People.* Sometimes one individual is needed, at other times three or four will be required. It is therefore necessary for you to gather a few individuals who are willing to take the time to help you with the practicing.

METHODS OF PRACTICING

There are four separate methods that are recommended for practicing:

1. Individual practice
2. One-on-one practice
3. Makeup oral board practice
4. Mock oral board practice

Individual Practice

Individual practice refers to those times when you are practicing all by yourself. No one is available to listen to you or make comments on your presentation. However, these sessions are extremely important and beneficial. They can be broken down into two types: silent practice and talking to yourself.

Silent Practice Silent practice refers to those times you take a few of the cards with you to ask yourself while waiting for a doctor's appointment, riding public transportation, relaxing at home, and so on. The procedure is for you to mentally ask yourself a question from one of the cards and mentally provide an answer. Silent practice is 95 percent mental. You do not need any tools except your cards.

Talking to Yourself Talk to yourself. Sit or stand before a mirror and look yourself in the eye (see Figure 10-1). It is extremely important that you learn to look a person in the eye when you are answering a question. Of all the people to whom you will be talking, you will have more difficulty looking yourself in the eye, keeping eye contact, and maintaining your thought process than you will with anyone else. If you can learn to keep your thought process and your organization intact while maintaining eye contact with yourself, you can do so with anyone. Three important questions you can ask yourself during this type of practice session are:

1. Why do you want the position?
2. What are the duties of the position?
3. What makes you think you are qualified for the position?

You will find that you will have a tendency to develop a monotone voice while talking to yourself, more so than when talking to others. Again, if you can maintain eye contact with yourself, keep your thoughts and organization intact, maintain your enthusiasm and voice modulation, and smile at yourself once in a while, you can do it with anyone.

Figure 10-1 Talking to yourself while making eye contact is excellent training.

Do not become discouraged during the first few times you try. It is demanding and challenging, but exceedingly rewarding when you master it.

All talking-to-yourself sessions should be voice recorded. Voice recording is necessary for you to determine if you are speaking in a monotone.

One-on-One Practice

If you have a friend you trust to provide an honest evaluation, you have a valuable resource. Give that individual your stack of cards and have him or her ask you the questions (see Figure 10-2). Have that person evaluate each of your answers. Listen carefully to what he or she says. You might think you have given a different presentation, but your friend is telling you exactly how you came across to him or her.

Do not be surprised if your best friend's evaluation is a little bit critical. Expect and encourage it. Pointing out errors and recommending methods for improvement is the only way you will benefit from these sessions.

These one-on-one sessions can be much more valuable if you have another friend who is willing to videotape it. An alternative to having another friend handle the videotaping would be to set up the video camera on a tripod and set the camera to record as you begin. If not, you will have to be satisfied with just listening to your voice recorder. Although the voice recorder gives you a good evaluation of your voice and presentation, the video recorder provides an excellent tool for evaluating your facial expressions and your demeanor.

Figure 10-2 A one-on-one practice session.

Makeup Oral Board Practice

This type of practice session should be reserved to be used after you have worked out many of the flaws that displayed themselves during your silent practice sessions and your one-on-one sessions. It will be necessary for you to gather at least three friends to represent the practice oral board. They will serve as members of the interviewing board (see Figure 10-3). If you prefer, you can have a fourth friend videotape the entire session. If you cannot arrange for this fourth person to videotape the session, you can set the camera on a tripod and turn it on as you begin. It is important that this type of session be videotaped.

Have the members of the practice board start out the session by asking you three questions: why you want the position, what makes you think you are qualified for the position, and what duties a person who occupies the position performs. After you have answered these three questions, the interviewers can ask questions by using the remainder of the cards.

Listen carefully to what the three individuals tell you during their evaluation of your presentation. A single friend can be extremely helpful in an evaluation, but it is much better to have more than one. When several people tell you that you came across in a certain manner, you are more likely to believe them than you would if only your best friend told you so.

Mock Oral Board Practice

Your final test and practice involves mock oral boards. While a practice oral board is valuable using friends, the most valuable is one set up for entry-level candidates in a fire

Figure 10-3 A makeup oral board practice session.

Figure 10-4 A mock oral board practice session.

station. If you are in this category, you should try to get a company officer or a higher officer to set up a practice oral board for you (see Figure 10-4). These officers have taken interviews and are familiar with the procedure. Ask the officer to duplicate as closely as possible the oral board to which he or she feels you may be subjected. Do not have the oral board use your cards. Have each member think of the questions to ask you. Also, do not limit your practice to one session. After the mock oral board has evaluated you and given you some advice on methods you can use to improve your presentation, go home and practice what you were told to do. Before you leave, see if you can set up an appointment for another session. The more mock orals you are given, the more likely you are to improve your performance.

Those individuals seeking promotion to company officer should follow the same procedure as those given for an entry-level firefighter. However, they should seek the help of chief officers in setting up the mock orals.

Tips for Success It has been found that individuals from the outside professional world who are involved in recruiting others for their own organizations can also be extremely helpful. This is particularly true if the individual is recruiting for management positions. Regardless of the position sought, the more practice an individual has, the more efficient he or she will become.

WHEN SHOULD I START PRACTICING?

Candidates normally ask when the practice sessions should commence. There is no set answer to this question. Much depends upon the need of the candidate. However, it is recommended that a three- to six-month period be considered. The period selected should commence just prior to your scheduled oral interview.

ENTRY-LEVEL FIREFIGHTER QUESTIONS

1. Tell us about yourself.
2. Why do you believe you are qualified to be a firefighter?
3. Describe to us the work schedule of a firefighter on our department.
4. What qualifications do you believe a prospective firefighter should have?
5. Which of the qualifications for firefighter do you believe is your strongest?
6. Which of the qualifications for firefighter do you believe is your weakest?
7. Describe to us the on-duty physical training program for the firefighters on our department.
8. Why do you want to be a firefighter?
9. Why do you think we should hire you rather than one of your competitors?
10. Tell us what you know about our fire department.
11. Why did you select our fire department to submit your application rather than another one?
12. What have you done to prepare yourself to be a firefighter?
13. Why do we want our candidates to be EMT-qualified?
14. A firefighter works a 24-hour shift on our department. Who prepares the meals for the firefighters?
15. What do you have to offer our department that no other candidate has?
16. What are your hobbies?
17. Which sports did you participate in while attending high school?
18. What made you decide to be a firefighter?
19. Have you submitted applications for a firefighter's position to any other city?
20. How do you think your previous employer would describe you?
21. Tell us about the different types of firefighting companies we have in our department and how many of each we have.
22. What is the name of the fire chief?
23. About how many alarms does our fire department receive in one year?
24. Describe the functions of an engine company.
25. Describe the functions of a truck company.

26. Approximately what is the population of our city?
27. What type of collar insignia does a fire captain wear?
28. What type of collar insignia does a chief officer wear?
29. If you are hired by our department, what do you expect to be doing five years from now?
30. If you are hired by our department, what position would you like to hold when you retire?
31. What word would you use to best describe your attitude?
32. Tell us about your education.
33. Tell us about the jobs you have had since leaving high school.
34. Give us three examples of incidents where you have been able to use some of the knowledge you have.
35. Give us three examples that demonstrate your ability to improvise.
36. Give us some examples of your ability to learn.
37. Give us some examples that illustrate you have the ability to think and act quickly.
38. What do you do to keep yourself in good physical condition?
39. Other than sports, give us some examples of your ability to work as a member of a team.
40. If you are offered a position on another fire department at the same time we offer you one, which one would you take, and why?
41. Give us some examples that illustrate you have the ability to get along with people.
42. Why should we hire you?
43. Give us some examples that illustrate you have the ability to present yourself well before a group.
44. Give us some examples of your ability to remain calm under stress.
45. Give us three examples of your ability to work with your hands.
46. Give us some examples to show that you are a prudent person.
47. Other than sports, give us some examples that illustrate you have endurance.
48. Give us some examples that illustrate you have good work habits.
49. Do you believe you are a hard working individual? If so, why?
50. Other than sports, what other evidence can you offer us to show that you have the physical strength and coordination to be a firefighter?
51. Can you give us any evidence that you are a self-disciplined individual?
52. What are your educational goals?
53. What are you going to do on your days off?
54. What relatives or friends do you have who are or have been firefighters?
55. What do your parents think about you becoming a firefighter?
56. What type of community activities have you participated in?

57. You are a rookie firefighter on probation. The members of the fire department go out on strike. Will you go with them?
58. What would you do if your company officer gave you an order you thought was legally or morally wrong?
59. After extinguishing a fire in a single family dwelling, the owner of the house brings out a pot of coffee and some cups for the firefighters. The department has a rule that no one is to accept gifts for performing their duties. A cup of coffee would really taste good to you. Would you take the coffee?
60. What would you do if your company officer orders you into a very smoky fire to make a rescue and there is no breathing apparatus available for you?
61. You are assigned to a truck company directly out of the drill tower. An hour after you report for duty, the company responds to a fire. At the fire the company officer orders you to raise a ladder in a manner that violates what you were taught. How would you handle this?
62. The company officer on the other shift where you work is a personal friend of yours. Your company officer starts asking you questions about the other officer's off-duty activities. How would you handle this?
63. Your company officer sends you into a structure to make a rescue. Inside you find a member of your company and your spouse. Both are unconscious and not breathing. You can take only one. The other one will die. Which one would you take?
64. You are at an emergency in which a fellow firefighter and several civilians have been injured. To whom would you give first attention?
65. If you become a firefighter, to what type of company would you like to be assigned? Why?
66. If the salary for a firefighter were 20 percent less than it is, would you still want the job? Why?
67. Your company officer tells you to get an axe and ventilate the roof. While attempting to carry out the officer's orders, the battalion chief stops you and tells you to get a $1\frac{1}{2}$-inch line and take it into the building. What would you do?
68. You are assigned to an engine house where you find you have a lot of spare time on your hands. How would you make use of this time?
69. Why do you feel that hiring you would improve our department?
70. Why have you chosen this department to start your fire service career?
71. Do you currently have any applications on file with other fire departments?
72. What kind of work have you done that you feel is directly related to the fire service?
73. What do you think a firefighter does other than put out fires?
74. What do you hope to be doing ten years from now?
75. You respond to a reported traffic accident. Upon arrival, you see a badly mutilated body in the street. A closer look indicates that the person is one of your best friends and that he or she is still alive. How would you react to this?

76. Do you have any objection to the department establishing boundaries within which you are required to live?
77. Do you believe you are mature enough for this job?
78. Why did you decide to apply for the position of firefighter rather than that of a police officer?

COMPANY OFFICER CANDIDATE QUESTIONS

1. Tell us something about yourself.
2. What makes you think you are qualified to be a company officer?
3. What qualifications do you believe a candidate who is applying for the position of company officer should have?
4. Describe to us exactly what duties a company officer performs.
5. What techniques do you use for solving problems?
6. What is your definition of leadership?
7. What is the biggest mistake you have made since joining the department?
8. What do you think will be your biggest problem as a company officer?
9. What methods do you plan to use to motivate members of your company?
10. You are assigned to a special project to cut the department budget by 10 percent. What recommendations would you make for accomplishing this assignment?
11. What recommendations do you have for increasing the productivity of the department?
12. What recommendations do you have for improving the department's emergency medical care procedures?
13. What do you think you will be doing ten years from now?
14. Of all the courses you have taken in college or in special seminars, which one do you believe has been most beneficial to you? Why?
15. What major improvements do you expect to be made in this department in the next five years?
16. What recommendations do you have for improving morale on the department?
17. Give us an example of a situation in which you displayed loyalty to the department.
18. What major items are in our budget this year that were not in last year's budget?
19. Tell us the characteristics of the worst company officer for whom you have worked.
20. Tell us the characteristics of the best company officer for whom you have worked.
21. What is our ultimate goal in the department?
22. How would you establish priorities in the management of your company?
23. What method are you going to use to evaluate the efficiency of our department?
24. How would you determine the training needs of your company?
25. What was the main article in the last fire service magazine you read?

26. What method would you use to determine the effectiveness of your training program?
27. Where do you want to work if you are promoted?
28. How do you feel that your promotion will help this department?
29. What do you believe is the biggest problem facing the fire service today?
30. In your opinion, what has been the biggest innovation in the fire service during the past five years?
31. How far ahead do you believe you should plan your drills and company activities?
32. What type of educational prerequisites do you believe should be established for the position of company officer?
33. List the courses you have taken to improve your computer skills.
34. What do you believe is the primary value of conducting fire prevention inspections at the company level?
35. What future problems do you anticipate for the fire service?
36. What type of terrorist activities do you believe the fire services in this country will eventually have to face?
37. What type of technical advances do you believe our department will use within the next five years?
38. What do you believe is the best indication that a firefighter is ready for a promotion?
39. Do you believe that our recruitment standards are being lowered or improved?
40. What books and magazines have you read in preparation for this examination?
41. What do you believe are some of the primary problems facing our department?
42. In what type of community activities are you currently involved?
43. You go into a smoky environment alone. Inside you find a member of your company, your mother, and the President of the United States. None of them are breathing. You can take only one. The other two will die. Which one would you take?
44. Other than a religious leader, who do you believe is the greatest leader that has ever lived on this earth?
45. What do you think of the leadership of your fire chief?
46. You see a member of your company pick up an expensive watch and put it into his turnout coat at a fire. How will you handle this?
47. You have been working at a fire from a factory that makes rope. You have picked up your equipment and are ready to return to quarters when a member of your crew tells you to take a good look inside the apparatus compartments. You do so and find them loaded with rope. How are you going to handle this?
48. The fire is out and your crew is taking a well-needed rest. The battalion chief drives up, takes a six-pack of beer out of his car and places it in front of your crew. He tells them to help themselves. How are you going to handle this?
49. One night after dinner you walk into the battalion chief's office and find him taking a drink of liquor. How are you going to handle this?

50. You enter the kitchen one night after dinner and find that all the members of your crew are playing poker. There is a large amount of money in the middle of the table. How are you going to handle this?
51. You find a member of your crew filling a five-gallon can with gasoline from the station supply. How do you intend to handle this?
52. A member of the other shift calls in sick. After being relieved, you go to the beach. You see the member who had called in sick riding a surfboard. How will you handle this?
53. You pull up behind a large gasoline truck at a stop signal on your day off. There is a large amount of gasoline spilling from the truck. You do not have a cell phone. How would you handle this?
54. You walk into the dormitory at bedtime and see a member of your company take a drink from a bottle and place the bottle under a pillow. How would you handle this?
55. You are the Incident Commander at a large industrial fire. Make up the details of the fire and give us your initial report to the dispatch office.
56. What system will you use for estimating the number of companies needed at a fire when you are the Incident Commander?
57. Tell us about the largest fire you have been to since joining the department.
58. You have been assigned the first female firefighter on our department. Tell us what problems you anticipate and what you are going to do about them.
59. You have been assigned the first African-American firefighter hired by our department. The firefighter is a male who is completely rejected by the members of your company. How are you going to handle this?

SUMMARY

Well, you've done it. Your journey is almost complete. Now the real work begins. Up to now, you have completed portions of your preparation to make a positive impression at an oral interview. Now it is time to put all the pieces together. It will take time and effort, but you can do it. The rest of the way is up to you. Good luck, and make your efforts pay off.

GLOSSARY

Age Discrimination Act This is an act that was passed by congress in 1967 to protect individuals between the age of 40 and 65 from discrimination in employment. In 1996, Congress passed a second amendment to the act that permanently allowed an exemption that permitted state and local governments to use age for hiring and retiring firefighters and law enforcement officers. This amendment is still in effect.

Appointing authority The person with the legal authority to hire and fire. This individual is normally the general manager of the hiring department. The Fire Chief is the appointing authority for the fire department.

Candidate Physical Ability Test (CPAT) A test that is available to all departments for use in the physical ability portion of the examination process. The test is quite extensive and physically demanding.

Certification For fire departments, certification means that the Civil Service Department is certifying to the Fire Chief that the candidates whose names are submitted to him or her have been examined in accordance with existing rules and have met the minimum standards for hiring.

Chronological resume A chronological resume highlights skills, training, and experience chronologically over a period of time.

Civil Service Act An act passed by congress in 1883 that was developed to create a government where the best-qualified individuals are hired.

Desirable qualifications Qualifications that are not essential in order to be hired but are desired by the examining jurisdiction.

Eligible List A list that indicates the names of candidates who have successfully passed the examination process and are eligible for hiring or promotion.

Examination bulletin A bulletin used by an examining jurisdiction to announce the date and time an examination will be given. The bulletin also lists examples of the duties to be performed and the minimum qualifications required for a potential candidate to compete in the examination process.

Filing (an application) Refers to submitting an application to take an examination.

Functional resume A functional resume highlights a variety of skills and qualifications that are relevant to the position sought. A candidate takes all the facts he or she wants to include in the resume and lists each one under one of four or five functions.

Gambling Betting on an uncertain outcome or playing a game for money or property.

Halo effect The tendency to grade an individual high on one factor based upon a favorable impression he or she made on a different factor.

Incident Management System (IMS) The incident management system has been developed to provide for the standardization of titles, duties, and terminology for the command and control of fires and other emergencies. It is also referred to as the Incident Command System (ICS).

Intent Used to determine whether an individual has taken something as a theft or because of poor judgment. If an individual takes something with the intent to employ it for his or her own personal use, it is theft.

International Association of Fire Chiefs (IAFC) An international association of fire chiefs that has been formed to advance the fire service profession.

International Association of Firefighters (IAFF) An international association of firefighters that has been formed to advance the fire service profession.

Interest card A card that can be filed with an examining jurisdiction that will be mailed to the candidate when an examination is to be given.

Merit system A system of filling government positions based upon the philosophy that the best-qualified person should be hired to fill a vacant position.

Minimum qualifications The minimum qualifications established by examining jurisdictions in order for an individual to compete in an examination process.

National Incident Management System (NIMS) A national emergency management system developed under the direction of the Secretary of Homeland Security. The objective for the development of the system was to provide a consistent nationwide template to enable all government, private-sectors, and non-governmental organizations to work together during natural disasters and emergencies, including acts of terrorism.

Non-stress board The board is normally very friendly and tries to place and keep a candidate at ease during the entire interview.

Open rule All names on the Eligible List are certified to the appointing authority when a single opening occurs. The appointing authority may hire any candidate from the list.

Personnel problems Situations in which a conflict exists and there are two or more possible solutions.

Physical ability test The practical portion of the examination process for entry-level firefighters. It consists of job-related activities that test a candidate's strength, agility, endurance, and coordination needed to perform the duties of a firefighter.

Polygraph test This test may be given individually to an entry-level firefighter candidate or as part of the background check. The objective of the test is to determine the honesty and credibility of a candidate.

Probationary period The last phase of the selection process. It is a period of time that individuals who are hired are required to serve before the appointment to the position becomes permanent. This period is referred to as the "working test period."

Problem The need to make a decision when there are two or more possible solutions.

Protest period A time frame in which a candidate may express unhappiness regarding perceived discrimination or prejudice, or a belief that the examination was conducted in a fraudulent manner.

Rapid Intervention Team (RIT) Personnel or a unit that is on standby in the vicinity of the command post for immediate deployment in the event of trapped, lost, or otherwise endangered firefighters. It is also referred to as rapid intervention crews (RIC) or firefighter assist search teams (FAST).

RECEO VS The seven basic strategies: rescue, exposures, confinement, extinguishment, overhaul, ventilation, and salvage.

Rule of One Only the top candidate is certified to the appointing authority when a single opening occurs. The appointing authority must hire the candidate or justify the rejection of the candidate in a written report to the Civil Service Commission.

Rule of Three Names are placed on the Eligible List according to candidates' final grade in the examination process. When an opening occurs, the top three names are certified to the appointing authority. The appointing authority hires the one he or she believes is best qualified.

Spoils system This system was based upon the concept that public positions are considered gifts of the controlling authority to be used for either personal or political purposes, or both.

Standard operational procedures (SOP) A system established by the fire administration to ensure that all firefighting units in the department operate using standard developed procedures.

Stress board The board attempts to evaluate a candidate's ability to operate under stress and may at times act mean and/or criticize a candidate to keep him or her under pressure or to increase the pressure.

Tenure After a hired employee completes the probationary period, he or she becomes a permanent member of the department. Tenure is a property right that cannot be taken away without due process of law.

Tier System An Eligible List is usually divided into three tiers. All names on the first tier are certified to the appointing authority when an opening occurs. The appointing authority may hire any candidate from the tier list.

Toastmasters/Toastmistresses Members of clubs that have been organized for the primary purpose of developing and improving members' public speaking ability.

INDEX

I

L

M

N

O

P

Q

R

S

T

U

V

W